Theory and Application of Ideal Wind Turbine
理想风力机理论与应用

Jiang Haibo　　Li Yanru　　Zhao Yunpeng

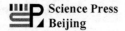

Title: Theory and Application of Ideal Wind Turbine
Authors: Jiang Haibo   Li Yanru   Zhao Yunpeng
Responsible Editors: Geng Jianye   Wu Zhou

图书在版编目(CIP)数据

理想风力机理论与应用 = Theory and Application of Ideal Wind Turbine: 英文/姜海波，李艳茹，赵云鹏著. —北京：科学出版社，2018
ISBN 978-7-03-056569-3

Ⅰ. ①理… Ⅱ. ①姜… ②李… ③赵… Ⅲ. ①风力发电机－研究－英文 Ⅳ. ①TM315
中国版本图书馆 CIP 数据核字(2018)第 028929 号

Published by Science Press
16 Donghuangchenggen North Street
Beijing 100717, P. R. China
Printed in Beijing

Copyright© 2018 by Science Press
ISBN 978-7-03-056569-3

All rights reserved. No part of this publication may be reproduced, stored in a retrieval system, or transmitted in any form or by any means, electronic, mechanical, photocopying, recording or otherwise, without the prior written permission of the copyright owner.

Preface

Human beings have thousands of years of history of design and use of wind turbine, but it remains a mystery for curious human beings how the most efficient ideal wind turbine is structured and what about its performance. It is one of the objectives of this book to explore the inner structure of ideal wind turbine and reveal its operating law.

Even the obtainment of all secrets of ideal wind turbine only satisfies the thirst for knowledge of human beings; this book further aims at providing guidance for the design of real wind turbine with ideal wind turbine theories.

This gives birth to a new method for aerodynamic design of wind turbine, namely the design of blade is simplified as making minimum modification to the structure of ideal blade, and the principle of modification, of course, is to make the blade sufficiently rigid and make it easy to produce, install and use with minimized loss of performance.

Fortunately, as the mathematical model of ideal blade (blade function) has been established and the practical blade function which is modified based on ideal blade function has also been built, now it is possible to obtain the three-dimensional graphs of blade rapidly by generating function graphs with mathematical software. This is known as functional design of blade with which it simply needs making adjustment to the structure of function or the value of constant to change the design and meanwhile the laborious manual design of stereoscopic graphics of blade will be replaced by instant computer-aided design, which means significant improvement of design efficiency. For example, compared with a manual drawing of a number of different sine curves, it is far simpler and more accurate, efficient and "once for all" to focus on constructing a sine function and generate graphs with software by adjusting amplitude, period and other constants; furthermore, this analytic function has more technical elements than that of its graphs and it is more easy to communicate, spread and improve than graphs.

Another benefit brought by blade function is that it enables rapid obtainment of the power, torque, lift and other performance parameters of wind turbine with analytic calculations, which is far more convenient and rapid than numerical calculation and experiment tests, and eliminates the necessity of generating graphs or physical model in advance. Although analytic calculation is not mature enough to substitute numerical calculation and experiment test at present, its performance prediction feature will contribute to reduced aimlessness in design and significant enhancement of design efficiency.

Centering on the above, this book will unfold the research by focusing on two aspects: One is the theory of ideal wind turbine, as illustrated in Chapters 2-6; The other is the application of this theory, namely the functional design of blade, as illustrated in Chapters 7-13.

In Chapter 1 (Overview), the contents, objectives, significance, approach and status of the research are outlined.

In Chapter 2 (Theoretical Basis and Basic Relational Expressions), force analysis is carried out for the wind turbine in design situations based on the Blade Element Momentum (BEM) Theory, also the basic relational expressions of relevant parameters are provided, so that the differential formula of the aerodynamic performance of blade element is obtained and the integral formula of aerodynamic performance of wind turbine is deduced, which serves as the theoretical basis of subsequent chapters.

In Chapter 3 (Structure of the Blade of Ideal Wind Turbine), the concepts of ideal blade and ideal wind turbine are put forward and it is demonstrated that the optimal angle of attack that provides the maximum power is the angle of attack with the largest lift drag ratio; besides, the structural features of ideal blade is explored, including ideal chord and ideal twist which serve as two key structural elements.

In Chapter 4 (Highest Performance of Ideal Wind Turbine), the computational formulas for the highest performance (with finite tip speed ratio) and limit performance (with infinite design tip speed ratio) of the power, torque, lift and trust of ideal wind turbine which is composed of ideal blades and operates in ideal fluid (with infinite lift drag ratio) are deduced.

In Chapter 5 (General Performance of Ideal Wind Turbine), the computational formulas for general performance (including power, torque, lift and thrust performance) of ideal wind turbine in real liquid (with both finite lift drag ratio and finite design tip speed ratio) are deduced, and the integral calculations (specific values) of power, torque, lift and thrust performance which correspond to any tip speed ratio and lift drag ratio are provided.

In Chapter 6 (Flat Airfoil Wind Turbine and Its Performance), the performance features of a special airfoil—flat airfoil are explored, including large angle of attack circumfluent lift and drag changes with angle of attack; for instance, the performance features of flat airfoil ideal wind turbine are further explored.

In Chapter 7 (Function Airfoil and Its Main Performance), the concept of function airfoil is put forward, the method of generating airfoil with function and the method of approximating existing airfoil with function are explored, and the pressure distribution, lift coefficient and other performance of the function airfoil are analytically computed.

In Chapter 8 (Simplified Analysis on Ideal Blade), the way how to simplify ideal blade is explored and numerical integration calculation and comparative analysis is carried out with respect to the cases in which efficiency is reduced due to a number of simplification methods respectively.

In Chapter 9 (Highest Performance of Practical Wind Turbine), the tip loss caused by limited number of blades of practical wind turbine is explored, the formulas for maximum power, torque, lift and thrust performance of practical wind turbine are deduced in consideration of tip loss, and the performance calculations corresponding to tip speed ratio and lift drag ratio are provided, which provides practical guidance for design of wind turbine.

In Chapter 10 (Practical Blade Structure Design), the structural design of practical blades is analyzed and explored, including the selection of airfoil, determination of design tip speed ratio, chord design and the method for correcting twist.

In Chapter 11 (Practical Wind Turbine Performance Calculation), the analytic calculation formulas of power, torque, lift and thrust performance of practical wind turbines are derived, and numerical integration samples are provided.

In Chapter 12 (Blade Function Design Methods), three subfunctions—airfoil, chord and twist of blade are used to construct blade function (mathematical model of blade) so that the purpose of designing graphs of blade by generating graphs with blade function is realized, and specific steps and examples of functional design of blade are provided.

In Chapter 13 (Steps of Designing Solid Model of Blade), the steps of designing a solid model of blade are provided and the problems that are likely to be encountered during blade modeling and corresponding solutions are discussed.

It should be noted that this book is intended to construct the basic framework of the technical theory system for functional design of blade, provide the general method and thought for functional design of blade and present the preliminary scheme and examples of functional design of blade; however, it is not expected that the research findings can be directly applied in actual design and production process of blade, and fulfillment of this objective requires continuous improvement and perfection by scholars, specialists and engineering technicians. This book will provide this target with a research basis, platform, framework, direction and a series of examples, laying a solid technical foundation to facilitate the subsequent research on the method for functional design of blade.

Publication of this book is sponsored by NSFC (National Natural Science Fundation of China) Program (No.: 51375489). The first draft of the book is a doctoral thesis which was written with the meticulous guidance of Cao Shuliang, a professor from Tsinghua University. In the research stage of the project and in the writing process of the book, enthusiastic assistance and support from Cheng Zhongqing, Wang Xiaojie, Mo Ji and Tan Lei was received. I, the author, would like to express my heartfelt thanks to all of them.

Some definitions and views in this book represent my own opinions and only serve as humble inspirations to stimulate exploration among the peers in order to promote progress of the research in this area. In addition, it is appreciated to give me comments and suggestions as the book contains shortcomings inevitably due to my limited capability.

Jiang Haibo
October 2017

Contents

Preface

Chapter 1 Overview ··· 1
    1.1 Contents of research ··· 1
    1.2 Objectives of research ··· 4
    1.3 Research status ·· 4
    1.4 Significance of research ·· 9
    1.5 Research approaches ··· 12

Chapter 2 Theoretical Basis and Basic Relational Expressions ······ 19
    2.1 Analysis of the force on blade elements in design conditions ····· 19
    2.2 Basic relations among parameters ···························· 20
    2.3 Differential formula of the pneumatic performance of blade element ········ 22
    2.4 Integral formula of the pneumatic performance of wind turbine ······ 24
    2.5 Summary of this chapter ··· 26

Chapter 3 Structure of the Blade of Ideal Wind Turbine ··········· 27
    3.1 The meaning of ideal blade ······································ 27
    3.2 Calculating the optimal angle of attack ···················· 28
    3.3 Calculation of ideal twist ·· 30
    3.4 Calculation of ideal chord ··· 31
    3.5 About ideal airfoil ·· 33
    3.6 The form of ideal blade ·· 33
    3.7 Summary of this chapter ··· 34

Chapter 4 Highest Performance of Ideal Wind Turbine ·············· 36
    4.1 Power performance and its limit ······························ 37
    4.2 Torque performance and its limit ····························· 40
    4.3 Lift performance and its limit ·································· 41
    4.4 Thrust performance and its limit ······························ 42
    4.5 Summary of this chapter ··· 43

Chapter 5 General Performance of Ideal Wind Turbine ············· 45
    5.1 General performance of power ································· 45

5.2 General performance of torque ·· 51
5.3 General performance of lift ·· 53
5.4 General performance of thrust ·· 54
5.5 Summary of this chapter ··· 54

Chapter 6 Flat Airfoil Wind Turbine and Its Performance ································ 56
6.1 Flat airfoil and performance estimation·· 56
    6.1.1 Brief introduction to the around flow lift at small angle of attack ············· 57
    6.1.2 Exploration of formula of around flow at large angle of attack ················ 58
    6.1.3 Drag coefficient of around flow at small angle of attack ························ 65
    6.1.4 The functional relationship between lift and drag································ 65
    6.1.5 The regularity of change of lift and drag ··· 67
    6.1.6 Lift drag ratio of flat airfoil ··· 68
6.2 Structure of ideal blade with flat airfoil··· 68
    6.2.1 The optimal angle of attack ·· 68
    6.2.2 Ideal twist ··· 69
    6.2.3 Ideal chord ··· 70
6.3 Performance of flat airfoil wind turbine ·· 70
    6.3.1 Power performance ··· 72
    6.3.2 Torque performance ·· 74
    6.3.3 Lift performance ··· 76
    6.3.4 Thrust performance ··· 77
    6.3.5 Startup performance ·· 78
6.4 Summary of this chapter ··· 80

Chapter 7 Function Airfoil and Its Main Performance ······································ 82
7.1 Function construction method for airfoil profile ······································ 82
    7.1.1 Simplification of Joukowsky airfoil expression ··································· 82
    7.1.2 Function construction method of general airfoil profile ························ 84
    7.1.3 Functional construction method of complex airfoil profile ···················· 86
    7.1.4 Functional construction method of smooth trailing edge airfoil profile ······· 87
    7.1.5 Function expression of wind turbine blade airfoil································ 89
    7.1.6 Airfoil parameter expression and airfoil ring view ······························· 91
7.2 Main performance calculation for function airfoil ···································· 97
    7.2.1 Function airfoil speed distribution ·· 97
    7.2.2 Function airfoil pressure distribution ·· 99
    7.2.3 Function airfoil lift coefficient calculation ······································· 102

7.3 Function airfoil and flow calculation application ··· 105
    7.3.1 Airfoil expression extension ··· 105
    7.3.2 Main purpose of airfoil analytic calculation ··· 107
    7.3.3 Limitation of airfoil expression ··· 108
7.4 Summary of the chapter ··· 109

Chapter 8 Simplified Analysis on Ideal Blade ··· 111

8.1 Blade chord and twist simplification ··· 111
    8.1.1 Simplification purpose and principle ··· 111
    8.1.2 Influence of drag coefficient on chord ··· 113
    8.1.3 Simplification method of chord curve ··· 114
    8.1.4 Blade lift and drag distribution ··· 117
    8.1.5 Blade attack angle and twist distribution ··· 119
    8.1.6 Outline example for simplified blade ··· 121
8.2 Performance of simplified blade wind turbine ··· 122
    8.2.1 Power performance ··· 122
    8.2.2 Torque performance ··· 124
    8.2.3 Lift performance ··· 126
    8.2.4 Thrust performance ··· 128
    8.2.5 Starting performance ··· 130
8.3 Impact analysis of simplification way on performance ··· 131
    8.3.1 Simplification way of blade chord ··· 131
    8.3.2 Impact of simplification way on performance ··· 132
8.4 Summary of this chapter ··· 133

Chapter 9 Highest Performance of Practical Wind Turbine ··· 134

9.1 Influence of blade tip loss on structure ··· 134
    9.1.1 Basic expression ··· 134
    9.1.2 Chord curve correction ··· 136
    9.1.3 Twist curve correction ··· 138
9.2 Highest performance calculation for practical wind turbine ··· 139
    9.2.1 Power performance calculation ··· 139
    9.2.2 Torque performance calculation ··· 143
    9.2.3 Lift performance calculation ··· 145
    9.2.4 Thrust performance calculation ··· 148
9.3 Summary of this chapter ··· 150

Chapter 10 Practical Blade Structure Design ··· 151

10.1 Determination of airfoil and optimum attack angle ··· 151

10.2 Determination of design tip speed ratio ········ 151
10.3 Straight chord design ········ 153
10.4 Twist correction and treatment ········ 155
10.5 Practical blade shape design ········ 157
10.6 Summary of this Chapter ········ 158

Chapter 11 Practical Wind Turbine Performance Calculation ········ 159
11.1 Power performance calculation ········ 159
11.2 Torque performance calculation ········ 161
11.3 Lift performance calculation ········ 164
11.4 Thrust performance calculation ········ 166
11.5 Summary of this Chapter ········ 167

Chapter 12 Blade Function Design Methods ········ 168
12.1 Mathematical model for blade of wind turbine ········ 169
    12.1.1 Coordinate transformation for airfoil function ········ 169
    12.1.2 Twist rotation transformation ········ 170
    12.1.3 Build blade mathematical model ········ 171
    12.1.4 Example for blade image generation ········ 172
12.2 Smooth transition among airfoils ········ 174
    12.2.1 Sine curve fairing method ········ 175
    12.2.2 Transition of cylinder and airfoil ········ 176
12.3 Functional design steps and examples ········ 179
    12.3.1 Blade functional design steps ········ 179
    12.3.2 Blade functional design example ········ 181
    12.3.3 Blade image generated by software ········ 191
    12.3.4 Solve wind turbine performance by software ········ 196
12.4 Summary of this Chapter ········ 198

Chapter 13 Steps of Designing Solid Model of Blade ········ 199
13.1 Design idea ········ 199
13.2 Insert section curves ········ 199
    13.2.1 Determine the surface equation ········ 199
    13.2.2 Insert section curves in SolidWorks ········ 203
13.3 Do lofting to obtain surface ········ 204
13.4 Repair and splicing of surface ········ 204
13.5 Summary of this chapter ········ 205

References ·· 206
Appendices ··· 210
　　Appendix Ⅰ　Key terms interpretation ·· 210
　　Appendix Ⅱ　Meanings of main symbols ··211
　　Appendix Ⅲ　Common relation index ··· 212

Chapter 1　Overview

1.1　Contents of research

The contents of research in this book may be divided into two parts: the theory of ideal wind turbine and the method for functional design of blade, and the latter is illustrated as the application of the former.

The theory of ideal wind turbine mainly deals with the structures and performance of ideal blades that can deliver the largest power, including mathematical models and function graphs of ideal wind turbine, as well as the requirements for solution of general and maximum power, torque, lift and thrust performance.

The method for functional design deals with how to adapt the ideal blade to practical applications, aiming to minimize the performance loss while satisfying the requirements for easy manufacturing, sufficient strength and so on in real environment. The exploration focuses on the mathematical model (which mainly involves the adaptions of three subfunctions—airfoil, chord and twist to practical use), performance calculation and method for functional design of real blades.

These two parts together constitute an independent and complete technical know-why system. See Table 1.1 for the main objects of research and their structural features, fluid environment and performance calculation system as well as the layout of chapters and sections.

Table 1.1　The main objects of research and their structural features, fluid environment and performance calculation system as well as the layout of chapters and sections

Objects of research	Fluid environment	Structural features	Performance calculation
Ideal blade (Chapter 3)	Ideal fluid	Ideal chord, ideal twist	
Ideal wind turbine (Chapters 4 and 5)	Ideal fluid	Made up of infinite number of ideal blades	Computation of the highest performance (Chapter 4)
	Real fluid		Computation of general performance (Chapter 5)
Flat airfoil (Chapter 6)	Ideal fluid / Real fluid	Both the thickness and curvature of airfoil are 0	Computation of lift and drag coefficients / Computation of the performance of flat wind turbine

Continued

Objects of research	Fluid environment	Structural features	Performance calculation
Functional airfoil (Chapter 7)	Real fluid	Generate airfoil profile with analytic function	Computation of pressure and lift coefficients
Practical blades (Chapters 8, 10 and 12)	Real fluid	Simplified analysis of the chord and twist (Chapter 8)	
	Real fluid	Structural design of practical blades (Chapter 10)	
	Real fluid	Mathematical model and shape design (Chapter 12)	
Practical wind turbine (Chapters 9 and 11)	Ideal fluid	Made up of finite number of practical blades	Computation of the highest performance (Chapter 9)
	Real fluid		Computation of general performance (Chapter 11)

To simplify the expressions, the mathematical models of airfoil and blade are referred to as airfoil function and blade function respectively herein, and the airfoil and blade expressed with these functions are called functional airfoil and functional blade for short respectively. Now the highlighted contents of research are further explained as follows.

1) Building the mathematical model of ideal blade

The research deals with how to build the mathematical model of ideal blade and the identified solution is to give the functional expression and its graph of ideal blade. The mathematical model of the blade of wind turbine is determined by airfoil, chord and the spanwise distribution function of twist; in other words, blade function consists of airfoil function, chord function and twist function. Therefore, the research in this section is further divided into four parts: airfoil function, chord function, twist function, and the method for constituting blade function with these three subfunctions.

2) Solving the highest and ultimate performance of wind turbine

In one of its significant applications, the mathematical model of ideal blade is used as the basis for exploring the performance of ideal blade. It is the ultimate goal of actual design of blade, and therefore is of great theoretical value. What is researched in this part includes deducing the power, torque, lift and thrust performance of the wind turbine that is composed of ideal blades in stable operation, the highest performance indicators related to tip speed ratio in ideal fluid environment and the ultimate performance that can be reached in various tip speed ratio conditions.

3) Building the mathematical model of real blade

Although the mathematical model of ideal blade is of great theoretical value, it cannot be applied directly in the practice of engineering and extensive improvements are needed, mainly because the complexity of the curves of ideal blade makes it difficult to manufacture the blades and it is incapable of satisfying the requirements for structural strength and the like in real fluid environment. Therefore, this part mainly deals with building the mathematical model of real blade on the basis of ideal blade, including exploring setting up blade root, correcting the tip, simplifying the chord curve, and smooth transition method for piecewise function, etc., and building functional mathematical model of real blade with analytic method. Just as that the function of ideal blade consists of three subfunctions, the function of real blade also consists of airfoil function, chord function and twist function, but each change to the structure will make these three subfunctions more complex and provide more details; therefore, this part focuses on establishing a one-to-one corresponding relationship between the structural change and functional expression.

4) Solving the performance based on the mathematical model of real blade

This part mainly explores the solution of performance of real blade. That is, how to establish a one-to-one corresponding relationship between the structural change and functional expression, laying the technical foundation for rational design of blade. The research on performance of airfoil involves laborious work, but the research results of both the existing experiments and analogue simulations provide the data about how the lift drag ratio changes with the angle of attack, while the present research only makes use of the lift drag ratio parameter of the airfoil and thus it is unnecessary to probe into more specific research process, which means significantly simplified research on the performance of blade. The integral formulas of power, torque, lift and thrust performance will be derived from integrating the micro-segment of mathematical model of real blade along wingspan segment; it is difficult to obtain analytical solution as the expression of integrand (function of real blade) is quite complex but it is certainly practical to obtain numerical solution which is sufficient for research on the performance of real blade.

5) Researching the method for functional design of blade

The method for functional design of blade is explored on the basis of the built mathematical model of real blade and performance computation, hoping to modify or optimize the design by adjusting the parameters of function, namely to obtain the appearance of blade instantly by generating function graphs at any time and obtain the power, torque, lift and thrust performance rapidly through automatic calculation of software. This part focuses on how to integrate the above research findings into a software module in an effective way to make the design easy, rapid and accurate, so as to broaden the engineering application vision of the research on project.

1.2 Objectives of research

The theoretical research of this book aims to explore the structure of the blade of ideal wind turbine on the basis of BEM Theory, express its shape with function graphs, and solve the general performance, highest performance and ultimate performance of ideal wind turbine in terms of power, torque, lift and thrust, thereby revealing the secrete and inherent law of the operation of wind turbine and establishing the theory of ideal wind turbine.

The engineering application target of this book is, based on the theory of ideal wind turbine, to adapt the ideal blade to practical use, build the mathematical model of real blade with analytical method, generate the function graphs of the shape and appearance of blade and compute its performance, and finally to realize the functional design of real blade.

The general objective of this book is to establish the basic framework of the technical theory system that integrates the theory of ideal wind turbine and the method for functional design of blade.

It should be noted that this book will provide the general method and thought for functional design of blade and present the scheme and examples of functional design of blade; however, it is not expected that the research findings can be directly applied in actual design and production process of blade, and fulfillment of this objective requires continuous improvement and perfection by scholars, specialists and engineering technicians. This book will provide this target with a research basis, platform, framework, direction and some examples, laying the technical foundation for facilitating the subsequent research on the method for functional design of blade.

1.3 Research status

Albert Betz, a German aerodynamicist, put forward that the maximum value of wind energy utilization factor (also called efficiency or power coefficient of wind turbine) was 16/27 (about 0.593) which was known as recognized Betz Limit in the field of wind turbine aerodynamics. Later research showed that Lanchester, a British scientist, had deduced the same limit value as early as in 1915, and it was reported in a piece of literature in Europe in 2007 that Joukowsky, a Russian scientist, obtained the same result independently in 1920. Therefore, strictly speaking, the maximum wind energy utilization coefficient should be called the Lanchester-Betz-Joukowsky Limit[1]. This part of history shows that the Betz Limit was demonstrated for three times in the same period. Since then, the human beings still fail to design a wind turbine that has wind energy utilization coefficient higher than such limit by far. The interaction of wind and rotating blades will give rise to induction factor both axially and tangentially and as a result reduces the velocity of incoming flow even in the optimum operating conditions. This phenomenon has been validated by modern aerodynamics and the practice of design and use of wind turbine, proving that the Betz Limit is insurmountable.

Recognition of the Betz Limit makes it meaningless to make too much effort in seeking an efficiency that surpasses the Betz Limit. Now the Betz Limit may be sought as a goal, and the design efficiency of wind turbine may be deemed as "ideal design" if it "approaches" the Betz Limit.

However, the practice in almost 100 years thereafter proves that the Betz Limit is still unattainable in that it is still impossible to reach an efficiency of modern wind turbine of above 0.5[2]. Is there another lower theoretical value? The research findings in this book prove the existence of this lower theoretical value: the theoretical efficiency of wind turbine is a function of design tip speed ratio and airfoil lift drag ratio, and the efficiency of wind turbine will be close to the Betz Limit only if both the lift drag ratio and tip speed ratio approach infinite values (Section 4.1), which shows that the reality is still far away from the Betz Limit, while it is the theoretical value of power coefficient relating to tip speed ratio and lift drag ratio that will provide a better guidance for design practice.

Does the torque performance of wind turbine have a theoretical limit in line with the power coefficient? The research findings in this book reveal that the torque coefficient of wind turbine is also a function of tip speed ratio and lift drag ratio, and there is a theoretical limit of the torque coefficient which is about 0.401 when the lift drag ratio is infinite and the lift drag ratio is 0.635 (Section 4.2); as for modern high-speed wind turbines which have a tip speed ratio of more than 6, their torque coefficients in instable operating conditions never exceed 0.1 (Section 5.2).

The obtainment of theoretical value of wind turbine performance under the conditions of finite tip speed ratio and lift-drag ratio gives rise to another leap in the depth of human being's knowledge of wind turbine. The theoretical derivation provides not only the highest performance objective that man-made wind turbine can reach (Chapter 4) but also the structure of the blade that helps reach such performance objective, namely the structure of ideal blade (Chapter 3), and the mathematical model of ideal blade are obtained (Section 12.1 for analytical expression and its graph).

Thus, the blade may be designed in a brand new way: make minimum modification to the ideal blade that has the highest performance to adapt it to practical use, so that it fulfills the requirements for processing, manufacturing, strength and so on in real environment while minimizes the reduction of performance. The structure and performance of ideal blade are referenced as the benchmark in each aspect in the process of adapting the blade to practical use, which helps effectively reduce the aimlessness in the design of blade and improve the design efficiency. In addition, based on the mathematical model of ideal blade, the process of adapting the blade to practical use may be expressed with function to realize the functional design of blade (Chapter 12). In my opinion, this method for functional design of blade will be the most significant development trend in the future.

The blade of wind turbine has a quite complex shape, and it is very difficult to build a mathematical model as the blade has three-dimensional, torsional and streamline airfoil, in particular the mathematical derivation process involves extreme difficulty. These restrictions hinder the research on the theory of ideal wind turbine and the method for functional design of blade. By far, it seems that no one has conducted systematic research on the whole blade with analytic method at home and abroad according to available reports, and the available literature only deal with airfoil, twist or chord separately.

It is the extremely difficulty mathematical expression that hinders building the mathematical model of blade. All airfoils are designed as streamline to minimize the drag, but it is rather difficult to express the streamline airfoil with simple analytic formula and to carry out analytic calculation of lift and drag. In the first half of the 20th century, the performance of airfoil was measured mostly through wind tunnel experiment, and airfoils and their performance were summarized in data sheets for future reference; with the development of computer technology and computational fluid dynamics in the second half of the 20th century, the design of airfoil was increasingly relying on the numerical simulation method.

The geometry of airfoil may be described using a variety of methods, and there are four dominant methods: discrete database method, shape parameterization method, shape function disturbance method and analytic function method. According to discrete database method, coordinates are used to describe the airfoil, and the airfoil lines may be obtained by threading the coordinate lattice with splines. Currently, discrete database method is the dominant method for describing the airfoil. According to shape parameterization method, a number of parameters are used to describe the geometric dimensions of each part of the airfoil, and the design variables have definite geometric meanings but analytical expression is not given[3]. According to shape function disturbance method, the linear superposition of the original airfoil and disturbance shape function determines the shape[4], and usually Hicks-Henne function is adopted as shape function[5]. This method is very dependent on the geometrical data of original airfoil, and if the shape of original airfoil is not smooth, the shape of design airfoil will not be smooth and this has direct impacts on the smoothness of the pressure coefficient curves. Analytic function method is such a method that one or more analytic functions are used to directly represent the shape of airfoil, such as American NACA 4-digit and 5-digit airfoil series and British C4 airfoil series that are expressed with polynomial in early years. But the change of parameters has great influence on the overall shape and it is hard to find an appropriate individual coding way for the genetic algorithm, so the polynomial method has almost been abandoned. Recently it is reported that series are used to express airfoil in some researches[6]. This method is suitable for expressing existing airfoils and may be used for optimizing the design to obtain final result directly, but slight adjustment to parameters is also apt to affect the whole.

It can be said that currently the design of airfoil is based on the means of expression that involve graphics or graphical data discretization and functional expression has not really been adopted. If the analytic function used to express airfoil has simple structure, definite geometric meaning of parameters and satisfying effect of fine adjustment, its combination with the powerful computing capability of computer will likely give birth to airfoils that have more outstanding performance. For this purpose, a method for how to construct complex airfoils with the analytic functions defined with mean line-thickness function of airfoil is explored herein(Section 7.1), and each of the parameters has a definite geometric meaning to facilitate fine adjustment of shape. With airfoil functions, it is also possible to solve the main performance parameters such as pressure distribution and lift coefficient with analytic calculation (Section 7.2), which is of great theoretic and practical significance for studying the relationship between the shape change and the performance change and for exploring the inherent law that governs the impacts of shape on performance.

As blade is critical to increase of the power coefficient of wind turbine, the design of its shape in particular the design of twist and chord is of great significance and has influence on many aspects such as work efficiency, production cost and safety; therefore, many scholars carry out their research with various methods from different perspectives. The research perspectives from which the shape of blade is designed are roughly divided into three categories: minimum cost target, maximum efficiency target and safety target. A literature puts forward a method for shape optimization design according to which the unit energy cost of the wind turbine is the optimization target and the chord, twist and relative thickness of blade are taken as design variables[7,8]. Recently, some scholars who attempt to optimize the shape of blade based on local wind velocity distribution with the target of maximum electric energy output[9,10] have made significant research progress. The most basic target is to explore the method for designing the shape of blade from the perspective of maximum output power[11,12]. In recent years, pneumatic elasticity, load characteristics and some other safety-related considerations are also introduced into the shape optimization design[13].

The research methods are also roughly divided into three methods: analytic method, experiment method and numerical method. The analytic methods mainly include conventional BEM Theory[14], the Glauert Vortex Theory which ignores airfoil drag and tip loss, the Wilson Method which amends the Glauert Theory, and the Complex Method according to which the design constrain condition is inequation and nonlinear constrain target is optimized and solved[15] and so on. The numerical method and experiment method require obtainment of the shape of blade in advance and then performance parameters are obtained through fluid calculation software or experiment to improve the design based on performance. These are rather specific design methods and most of them emphasize practicality, which is beneficial to actual design of blade, but they seldom directly contribute to the building of mathematical model and functional design of blade.

Analysis from the perspective of mechanical design reveals that, from the sole point of view of drawing the shape of blade without performance computation, currently all of the three-dimensional shaping design methods that are based on various professional graphics software share the characteristic that the discrete coordinates of airfoil are entered into large professional graphics software like Pro/E, SolidWorks and UG after coordinate transformation to generate three-dimensional graphs of blade after manual smoothing or modification[16,17]. Most of the professional graphic software are capable of parameterization design and it is effortless to re-set the structural and dimensional parameters to facilitate modification and design.

But for drawing of three-dimensional graphs, there is no such a method that is more accurate and rapid than function graph generation method by far. Once there is functional expression (such as the functional expression of an ellipsoid), drawing of three-dimensional graphs can be completed instantly and modification of design will be nothing but adjustment of parameters. The bottleneck of such functional design mainly lies in that it is difficult to give functional expressions that have complex structures, and this gives birth to professional graphics software which omits the process of constructing complex functional expressions and provides the "what you draw is what you get" design method, eliminating the necessity of constructing complex functions which is replaced by manual drawing of complex graphs. But there is a new problem: the design of each unit of each individual is relatively independent from that of others and each design has to go through all complex steps from introduction of airfoil coordinates through smoothing the splines to stretching, torsion and so on, involving considerable rehandling and heavy work load. Although the function of blade is rather complex, the laborious work involved in drawing graphs will be performed by mathematical software for instant completion once the function is built, and thus all repetitive work will become nothing but adjusting the parameters of the function to alter the design. Independent and separate working is far less beneficial than focusing on building a complex but once and for all function which facilitates inheritance; therefore, in my opinion, the blade functional design method contributes to significantly reduced design efforts and minimized rehandling in line with the trend of the times. The functional expression of real blade preliminarily established in this book (Section 12.3) is waiting for continuous improvement by specialists, scholars and engineering technicians and it will certainly have promising development and application opportunities in the future.

1.4　Significance of research

The wind turbine operating in ideal fluid environment is certain to have the most ideal blade structure and the highest performance. The most ideal structure and performance of blade will provide an objective benchmark for judging the reasonableness of the design, and it indicates the direction for actual design of blade, in other words, the design will be a straightforward shortcut if the design of blade is based on the most ideal structure and performance with minimum performance loss and satisfies the requirements for strength, vibration and other indicators through structural modification, which facilitates designing real blades of which the structure and performance is close to structure and performance of the most ideal blade, and thus such blades will have relatively high performance and will be rather straightforward than aimless design.

In this book, the blade that delivers the maximum power is called ideal blade. The research findings herein have proven that the chord function and twist function of ideal blade have ideal distribution pattern in spanwise direction (Chapter 3), while the airfoil does not have ideal pattern (but its structural features may be described with lift drag ratio). Therefore, the theory of this ideal blade mainly deals with the structure and performance of ideal blade, and functional design method will deal with exploring the structure and performance of ideal blade after adapting it to practical use. As the contour structures of both ideal blade and real blade are generated by the graphs of analytic functions and all performance values are computed with the integrals of analytic functions, such design method is called functional blade design method or may be called analytic function-based blade design method in this book.

Conventional blade design consists of two parts: contour structure design and performance computation. Blade structure design is realized mainly through manual modeling with the help of large professional graphics software such as Pro/E, SolidWorks and UG after elongation, torsion, setting out, shelling and some other steps (it is impractical to express the shape of blade with the graphs of analytic function), and a full design cycle takes several days or weeks. After that the designed contour structure will be imported into fluid software like Fluent, CFX and others for calculation of performance values, which takes several hours, several days or weeks depending on the capacity of computer and meshing. These two phases of blade design are mutually dependent and the modification of design is almost a repetition of this process from the very beginning until the end, which means considerable labor, costs and resources and further improvement of design efficiency are almost impossible.

Compared with conventional blade modeling and numerical calculation methods, the practice of generating graphs of blade and carrying out analytic calculations of performance with blade functional blade design method enjoys a number of advantages. ①As the mathematical model framework of blade has been obtained when the research described herein is being conducted, it is only needed to plug three subfunctions—airfoil, chord and twist into the framework to obtain corresponding blade function. The design focuses on rational determination of subfunctions and in general elementary analytic geometry is sufficient for solving most of the problems. Even if the blade function is rather complex, it is far simpler to build expressions than to draw graphics manually based on complex expressions, while the latter should have been the basic function of mathematical software. ②Once blade function has been built, structural design becomes nothing but to generate blade graphs with blade function instantly with the help of computer; the performance calculation according to analytic method is numerical integration of the performance of blade function in spanwise direction and general mathematical software only takes several seconds to finish this process. Either structural design or performance calculation may take place first, and it is even possible to carry out performance calculation without generating contour structure, which means a leap forward of design efficiency. ③Once blade function has been built, the modification of design will become nothing but adjusting the parameter values in three subfunctions, and this is much easier than re-drawing graphics and re-performing calculations with large graphics software, which is extremely beneficial to design analysis. The adjustment of parameter values is quite simple because each of the variables of the blade function has a simple but definite geometric meaning. ④As spatial position parameters are determined by functional formulas, it is easy to obtain the accurate coordinates of any point on the surface of blade with the minimum probability of error and thus it becomes unnecessary to make a huge number of discrete lattice databases. It is easy to drive CNC (Computer numerical control) equipment with function formulas, which facilitates realizing automation, high efficiency and precision in processing and manufacturing process of blade of its model with very promising development potential. ⑤By establishing blade function framework platform, the technical professionals in this field will be capable of concentrating on research of the specific form of subfunctions and perfecting blade functions with joint efforts. Compared with the mode in which the technicians in different units explore their own drawing skills separately, such mechanism enjoys scale advantages and thus is more likely to promote the technical development of blade design.

Here is a simple example to demonstrate the above five advantages: manual drawing of a number of different sinusoidal curves is less simple, efficient and once for all than concentrating on building a sinusoidal function and generating a variety of graphs of sinusoidal function automatically by adjusting the amplitude, cycle and some other parameters. Although the "complexity" of this sinusoidal function is beyond the comprehension of "ordinary people", a little bit training will enable it to adjust parameter values and generate different graphs, and this analytic function has more technical elements than that of its graphs and it is more easy to communicate, spread and improve than graphs.

This book is not intended to make partial improvements on the basis of existing blade design methods; instead, it aims at establishing the basic framework of a new set of complete theory system for functional design of blades. This book is not intended to override the existing methods, but to provide a completely different option for the designers of blades.

1.5 Research approaches

With respect to the research approaches for complex problems, the research approaches may be categorized as three dominant methods—analytic method, experiment method and simulation method by the nature of relative independence and systematicness[18].

Analytic method is such a mathematical method that variables are used to express the attribute parameters of natural phenomenon and formulas are used to express the interaction relations among the attribute parameters. Analytic method is the most basic research method and the foundation of the research on natural science theories. Analytic formulas indicate clearly the interaction relations among the variables and show clearly the trend and degree of the influence of the change of any independent variable on dependent variables. If the interaction relation within or among natural phenomena can be expressed with analytic formulas, this indicates that humans have grasped its law.

Experiment method refers to such a method according to which some natural phenomena are recurrent under controlled conditions and instruments are used to measure the attribute parameters of recurrent natural phenomena. Experiment method only deals with natural phenomena objectively with measurement of results rather than reflects the mutual interactions among the parameters within natural phenomena; therefore, compared with other methods, the results obtained with this method are the closest to reality.

Simulation method refers to such a method that computer is used to carry out simulated recurrence and numerical calculations for some phenomena[19]. The object of simulation method is a complex system which requires building mathematical model, but usually the model is differential or integral equation(set) that is difficult to be resolved. This method is mainly characterized by the possibility of numerical solution of this equation(set), so sometimes it is also called numerical method. Simulation method is typically defined as "simulation is model-based experiment" in preliminary stage[20], and in recent development reports it is emphasized that "simulation is a model-based activity", including three basic activities on the research objective—test, analysis and evaluation[21].

For simple problems, humans have adopted numerous analytic methods in the research such as Newton's Second Law and theorem of momentum. For complex problems, analytic method encounters considerable difficulties. For example, the pneumatic calculations of the blades of wind turbine involve three-dimension, torsion, rotation and other complex conditions, and also involve the interactions among numerous variables including chord, span, twist, thickness, degree of curvature, airflow induction speed and air density, etc., and therefore the research is exposed to considerable difficulties.

To overcome these difficulties brought by complex problems, people adopt experiments as auxiliary research means, such as wind tunnel experiment. But experiment methods require high costs and long cycles. Experiment methods are only effective for forward problems, namely the structure and shape must be given in advance in order to solve the aerodynamic performance, while it is impossible to deduce the parameters of the structure and shape based on the optimum performance, and namely they are incapable of solving inverse problems.

With the development of computer technology, computational fluid dynamics and simulation software contribute a lot to solving the calculations of complex structures, and numerical calculation also replaces part of the functions of experiment. But there are still a lot of shortcomings. For example, compared with experiment method, the calculations of airfoil around flow contain considerable errors or occupy excessive computer resource, and only capable of solving forward problems rather than inverse problems.

It is worth noting that, for solvable problems, analytic method is capable of performing analysis, comparison and research of the performance of various structures and solving inverse problems. For example, it is easy to judge the influence of the change of any parameter on the overall performance, and thus it is possible to design the structural parameter values more reasonably and even possible to deduce the optimal structural parameter reversely. This is beyond the capability of both experiment method and simulation method. Although experiment method and simulation method deliver overall results, it is quite difficult to determine the degree of influence of a parameter on the results. It is needed to spend plenty of time to set different combinations of numerous parameters for testing or numerical calculation, and then pick the best result out of the limited results.

All these three research methods have their own distinctive characteristics, as shown in Table 1.2.

Table 1.2 Comparison of the characteristics of analytic methods, experiment methods and simulation methods

Compared items	Analytic methods	Experiment methods	Simulation methods
Object of research	Existing system, hypothetical system	Existing system	Existing system, hypothetical system
Purpose of research	Revealing, recurrence, prediction	Demonstration, verification or determining the conclusion	Demonstration, computation, or enhancing the performance of system
Basic approach	Abstraction, simplification, deduction	Simulation, recurrence, measurement	Building model equation and numerical solution equation
Theoretical or technical basis	Formal logic, either deductive or inductive	Theory-guided experiment, carried out with inductive method	Formal logic and Boolean logic, either deductive or inductive
Main problems that may be encountered in research process	Ambiguity, difficult deduction, and impossibility of obtaining analytical solution	It is difficult to reproduce the environment, and measurement is difficult	Inaccurate model, non-scientific algorithm and time-consuming calculation
Research findings	Functional expression or value obtained with expression	Measured discrete data or discrete database	Discrete data or its organic integration obtained through numerical calculation
Accuracy of findings	Determined by the degree of simplification, and may be either very high or very low	High	Relatively high
Reveal the internal regularity	Strong	Weak	Weak
The capability of solving inverse problems	Strong (extreme values are solvable)	Weak (such as local data fitting)	Weak (such as local screening method)

Each method has its own advantages and disadvantages, and in fact these three methods are penetrating into and promoting the development of each other in the course of research. Taking the analytic method for instance, it always needs experiment to measure attribute parameters and to verify the deduced results; if the expression that is deemed as the final result is quite complex, numeric method has to be sought to obtain numerical solution. Simulation method also needs analytic method to build differential equation and other mathematical models, and the numerical calculation results have to be compared with experiment results. Experiment method must be done in controlled conditions under the guidance of existing theories. It is strongly purposive: either to explore the important problems identified in the course of research on theories through experiment to conclude the general law, so as to provide the basis and reasonable assumption for the analysis conducted with analytic method and simulation method, or to verify or confirm the deduced results or the results of numerical calculation.

After a comprehensive assessment based on the above analysis, it is decided to explore the functional design of blade of wind turbine mainly with analytic method in this book.

According to analytic method, symbols are borrowed to reflect attributes of things and formulas are used to reflect the interactions of the attributes of things. The main difficulties encountered in the course of research lie in whether:

(1) The essential attributes of things can be expressed with symbols accurately.

(2) For complex problems, the essential attributes of things can be expressed with simplified expressions. For example, for shape attributes of an object, a circle can be expressed in a simple expression, but it is difficult to express the shape of airfoil in analytic expression.

(3) For a complex problem, there are excessive symbols or expressions that reflect the essential attributes of things, which makes it difficult to do calculation and derivation.

(4) There are axioms, theorems, laws or theories that interrelate these symbols or expression, or it is possible to put forward scientific assumptions by observing the internal laws of things to facilitate subsequent derivation.

(5) The derivation process is cumbersome, and a simple method can be found or software is available for assistance.

(6) In case difficulties are encountered in the above process, the problems can be

simplified to obtain simplified expressions that is capable of qualitative analysis.

With the passage of time, some of the difficulties encountered when adopting analytic method can be overcome by introduction of new graphical expressions, discovery of new theorems and the progress of mathematical software, etc., all of which might help overcome such difficulties.

Analytic methods provide different specific steps and approaches for different problems. For complex problems, it is more reasonable to carry out the research based on existing theories or data rather than start from the very beginning. The steps and approaches of analytic method for analyzing the performance of wind turbine for example are illustrated as follows.

1) Analyze the problem and judge the state of flow

First of all, the environment of the problem must be clarified. There are three airfoils around flow types: small angle of attack around flow, large angle of attack separating flow and stall area flow between them. The flow is small angle of attack around flow when the wind turbine is running in the optimum operating state. Large angle of attack separating flow is imposed on the blade at the moment when the wind turbine is just to be started. Sometimes it becomes rather tough to judge the state of flow, and even the same blade is experiencing all the above states of flow simultaneously. For example, the blades are sure to experience a process of change from large angle of attack state to small angle of attack state during startup of the wind turbine; in this process, the tip of blade will first be in the state of small angle of attack and the blade root will be the last to be in the state of small attack angle, while the middle of blade is usually left in stall and transition state.

2) Build the analytic expressions of airfoil lift and drag

It is the basis of carrying out analytic calculation of blades to build the expressions of airfoil lift and drag. For a given airfoil, it is needed to draw the curves which show how the lift and drag change with the change of angle of attack by referring to airfoil data, and then build the functional expression for how the lift and drag change with the change of angle of attack through mechanism study, regression analysis and other approaches, or provide the specific value of the maximum lift drag ratio. It should be noted that, if the built expression is excessively complex, the subsequent analytical calculation will be very difficult and thus analytic calculation will become meaningless. Flat airfoil is the simplified version of streamline airfoil and its formula showing how the lift and drag coefficients change with the change of angle of attack is relatively simpler, so it could be used as standby approach for simplifying the calculation to facilitate quantitative analysis.

3) Determine the induction speed

The wind velocity flow through blades is less than the velocity of free incoming flow, and the reduced portion of wind velocity is known as induction speed. The induction speed is generated by the interference of rotating blades on the free incoming flow; for horizontal axis wind turbine, induction speed is generated both axially and circumferentially. For example, when the wind turbine is in stable operation, the axial induction speed may reach up to one third of that of incoming flow, while the circumferential velocity changes a lot along the radius direction and the closer the point is to the root, the higher the induction speed will be. Only after correcting the velocity of free incoming flow can calculation be performed by using lift and drag formulas; otherwise, there will be enormous errors or even the calculations become useless.

4) Determine the inflow angle, angle of attack and twist

Although these three parameters are inter-related, they all represent the essential attributes of the state of blade, and thus accurate relational expression for calculation should be provided for them. As both lift and drag are functions of the angle of attack, angle of attack is one of the most important parameters. Being different from twist, angle of attack is not a structural parameter of blade, so it is always in dynamic change. After the wind turbine starts, the blades are rotating but the free incoming flow does not act on the blade axially, instead it acts on the blades along the direction of the resultant velocity of equivalent linear velocity which is opposite to the rotating direction of blades; the inflow angle is the angle between the direction of resultant velocity and the rotation plane of blades and such angle is changing dynamically as the rotating speed changes. But it is equal to the sum of the twist and angle of attack and thus the angle of attack can be calculated accordingly, which provides the preconditions for computing lift and drag.

5) Determine the chord changes

As lift and drag are proportional to the size of chord, they are also key parameters. Chord changes in wingspan direction and the formula for expressing chord change may be determined using the BEM Theory and the lift and drag formulas for the optimal angle of attack before subsequent integration can be performed.

6) Perform integration for solving performance

This step deals with derivation of the parameters or expressions obtained in the above steps which merely illustrate the calculation of the micro-segments of blades (blade elements), while the integration here will apply the force imposed on wind elements to the whole blade. Integration of the lift and drag of wind turbine or the torque, power and other parameters generated by them will provide the performance of the entire blade or the wind turbine that is composed of the blades. If the integrand expression is too complex, it is optional to use symbolic integration software to facilitate the integration or finally obtain numerical calculations.

7) Performance analysis

The performance can be analyzed once the analytic expression of performance function is obtained through integration. If the expression is very complex, it is optional to draw function graphs and observe the changing trend of performance curves. To be specific: draw the curves that reflect the relation between function and a key independent variable (the value of other independent variable shall be fixed), for example, draw the curve that reflects how the performance changes with the change of tip speed ratio while fixing the frictional resistance coefficient and other independent variables, and then analyze the performance by observing the trend of curve, providing analysis- based input for further improvement of performance through reasonable adjustment of the chord, twist and other structural parameters of the blade.

The research herein is mainly based on analytic method, but some achievements of experiment method and simulation method are also adopted. For example, the data of airfoil around flow experiment and airfoil pneumatic performance value analysis software is used for help.

Chapter 2 Theoretical Basis and Basic Relational Expressions

This chapter is mainly intended to carry out analysis of the force for the blades running in stable conditions (design conditions), explore the interaction between wind and blade, provide the basic formulas for expressing relations among parameters, and derive the pneumatic performance differential formula of blade elements and the pneumatic performance integral formula of the wind turbine. This content exists as the theoretical basis of studying the structures and performance of the blades of wind turbine and the formulas contained herein are repeatedly referenced in the following chapters and sections.

2.1 Analysis of the force on blade elements in design conditions

Design conditions refer to that the wind turbine is in stable operation or the optimal operating conditions. This section is intended to provide the basic formulas of design conditions for following chapters. These formulas are derived with blade element theory, namely the spanwise micro-segments of blade are deemed as blade elements and the lift and drag generated by such elements is integrated to obtain the integral expression of the overall performance of blades and the wind turbine.

Before researching the design of blades of wind turbine, let's briefly review the action of wind and the force on blades (Figures 2.1 and 2.2).

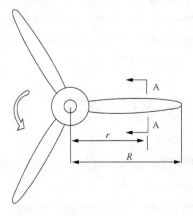

Figure 2.1 Diagram of tail wind rotating impeller of wind turbine

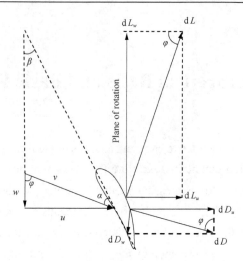

Figure 2.2 Blade element wind velocity state and force analysis of A-A section in Figure 2.1

Cutting the blade with a cylinder which is concentric with the rotating axis (A-A section in Figure 2.1 for the planning point), and take a micro-segment (blade element) d r at radius r to do performance analysis, as shown in Figure 2.2. Supposing the angular velocity of rotating wind turbine is ω, micro-segment d r moves upwards along the rotating plane, so there is a dead wind relative velocity $W=\omega r$. The resultant velocity of W and tangential induction speed bW is $w=(1+b)W$. The resultant velocity $u=(1-a)U$ of the absolute velocity U of incoming flow from an infinite point and axial induction speed aU is the velocity of the wind passing through the blade axially, the resultant of u and w imposes lift and drag on blade and the angle of attack of velocity v is α. Being perpendicular to the resultant velocity v, the lift d L may be decomposed into circumferential component d L_w and horizontal component d L_u; the direction of drag d D is consistent with that of v, and it may be decomposed into circumferential component d D_w and horizontal component d D_u. Spanwise (radial) integration of these forces is the resultant force imposed on blade.

2.2 Basic relations among parameters

During the design of wind turbine, it should be assumed that the wind turbine is running in the most ideal state (design situation), and the most important thing at this time is to determine chord and twist.

Design conditions are different from operating conditions. In the case of operating conditions, the chord and twist have been determined, and what is to be explored is how the performance of wind turbine changes with the change of tip speed ratio within a broad range, namely chord and twist are given, tip speed ratio is independent variable and performance parameters are dependent variables. Design conditions refer to the optimal operating conditions or stable operation conditions, and in such conditions the design tip speed ratio is predefined and is a constant, and what is to be explored is how to design the chord and twist to enhance the performance of wind turbine with the given tip speed ratio. It should be noted that, being different from operating conditions, the design conditions involve a stable value of speed induction factor, so the process of iterative solution of velocity induction factor may be omitted.

According to Figure 2.2, inflow angle φ is determined by the formula below:

$$\tan\varphi = \frac{u}{w} = \frac{(1-a)U}{(1+b)\omega r} = \frac{1-a}{1+b}\frac{1}{\lambda} \quad (2.1)$$

In stable operation state, the axial velocity induction factor a and circumferential speed induction factor b are stable values[22]:

$$a = \frac{1}{3}, \quad b = \frac{a(1-a)}{\lambda^2} = \frac{2}{9\lambda^2} \quad (2.2)$$

Therefore

$$\tan\varphi = \frac{1-a}{1+b}\frac{1}{\lambda} = \frac{1-a}{1+\frac{a(1-a)}{\lambda^2}}\frac{1}{\lambda} = \frac{6\lambda}{9\lambda^2+2} = \frac{6\lambda_t(r/R)}{9\lambda_t^2(r/R)^2+2} \quad (2.3)$$

where R is the length of blade, namely the distance from the tip to the center of rotating shaft of wind turbine; λ is the ratio of the tangential linear velocity W at r and the absolute velocity U of the incoming flow from an indefinite point, known as linear velocity ratio; λ_t is tip linear velocity ratio, known as tip speed ratio. See Appendix II for the meanings of other parameter symbols which are not specified in this book.

The maximum inflow angle is about 35.3°. The curve of inflow angle when wind turbine is in stable operation is given and does not change with the change of twist β (the angle between airfoil chord and rotating plane), but according to Figure 2.2, the twist and the angle of attack are mutually complementary and their sum equals to the inflow angle:

$$\varphi(r) = \beta(r) + \alpha(r) \quad (2.4)$$

It can be derived from Figure 2.2 that the inflow velocity is

$$v = \sqrt{w^2 + u^2} = \sqrt{(1+b)^2 W^2 + (1-a)^2 U^2}$$

$$= U\sqrt{\left(1+\frac{2}{9\lambda^2}\right)^2 \lambda^2 + \left(1-\frac{1}{3}\right)^2} \quad (2.5)$$

$$= U\sqrt{\left(\lambda+\frac{2}{9\lambda}\right)^2 + \left(\frac{2}{3}\right)^2}$$

The sine and cosine expressions of inflow angle can also be derived thereby

$$\sin\varphi = \frac{u}{v} = \frac{\left(1-\frac{1}{3}\right)U}{U\sqrt{\left(\lambda+\frac{2}{9\lambda}\right)^2 + \left(\frac{2}{3}\right)^2}} = \frac{\frac{2}{3}}{\sqrt{\left(\lambda+\frac{2}{9\lambda}\right)^2 + \left(\frac{2}{3}\right)^2}} \quad (2.6)$$

$$\cos\varphi = \frac{w}{v} = \frac{\left(1+\frac{2}{9\lambda^2}\right)\omega r}{U\sqrt{\left(\lambda+\frac{2}{9\lambda}\right)^2 + \left(\frac{2}{3}\right)^2}} = \frac{\lambda+\frac{2}{9\lambda}}{\sqrt{\left(\lambda+\frac{2}{9\lambda}\right)^2 + \left(\frac{2}{3}\right)^2}} \quad (2.7)$$

These basic relational formulas enable the derivation of the analytic calculation formulas of pneumatic performance of the blades which are in stable operation.

2.3 Differential formula of the pneumatic performance of blade element

Now preconditions are fulfilled to derive the formulas of thrust, lift, torque and power of blade element as per the blade element theory[23] based on the aforesaid formulas.

The lift of blade element is

$$dL = \frac{1}{2}\rho v^2 C \cdot C_L \, dr = \frac{1}{2}\rho \left[U\sqrt{\left(\lambda+\frac{2}{9\lambda}\right)^2 + \left(\frac{2}{3}\right)^2} \right]^2 C \cdot C_L \, dr$$

$$= \frac{1}{2}\rho U^2 C C_L \left[\left(\lambda+\frac{2}{9\lambda}\right)^2 + \left(\frac{2}{3}\right)^2 \right] dr \quad (2.8)$$

where, ρ is air density.

The drag of blade element is

$$dD = \frac{1}{2}\rho v^2 C \cdot C_D \, dr = \frac{1}{2}\rho \left[U\sqrt{\left(\lambda+\frac{2}{9\lambda}\right)^2 + \left(\frac{2}{3}\right)^2} \right]^2 C \cdot C_D \, dr$$

$$= \frac{1}{2}\rho U^2 C C_D \left[\left(\lambda+\frac{2}{9\lambda}\right)^2 + \left(\frac{2}{3}\right)^2 \right] dr$$

(2.9)

According to Figure 2.2, the total axial thrust of blade element may be derived with Formulas (2.6) - (2.9).

$$dT = dL_u + dD_u = dL\cos\varphi + dD\sin\varphi$$

$$= \frac{1}{2}\rho U^2 C C_L \left[\left(\lambda+\frac{2}{9\lambda}\right)^2 + \left(\frac{2}{3}\right)^2 \right] \frac{\lambda+\frac{2}{9\lambda}}{\sqrt{\left(\lambda+\frac{2}{9\lambda}\right)^2 + \left(\frac{2}{3}\right)^2}} dr$$

$$+ \frac{1}{2}\rho U^2 C C_D \left[\left(\lambda+\frac{2}{9\lambda}\right)^2 + \left(\frac{2}{3}\right)^2 \right] \frac{\frac{2}{3}}{\sqrt{\left(\lambda+\frac{2}{9\lambda}\right)^2 + \left(\frac{2}{3}\right)^2}} dr$$

(2.10)

$$= \frac{1}{2}\rho U^2 C \left[\left(\lambda+\frac{2}{9\lambda}\right) C_L + \frac{2}{3} C_D \right] \sqrt{\left(\lambda+\frac{2}{9\lambda}\right)^2 + \left(\frac{2}{3}\right)^2} \, dr$$

According to Figure 2.2, the total circumferential lift of blade element may be derived with Formulas (2.6) - (2.9).

$$dF = dL_w - dD_w = dL\sin\varphi - dD\cos\varphi$$

$$= \frac{1}{2}\rho U^2 C C_L \left[\left(\lambda+\frac{2}{9\lambda}\right)^2 + \left(\frac{2}{3}\right)^2 \right] \frac{\frac{2}{3}}{\sqrt{\left(\lambda+\frac{2}{9\lambda}\right)^2 + \left(\frac{2}{3}\right)^2}} dr$$

$$- \frac{1}{2}\rho U^2 C C_D \left[\left(\lambda+\frac{2}{9\lambda}\right)^2 + \left(\frac{2}{3}\right)^2 \right] \frac{\lambda+\frac{2}{9\lambda}}{\sqrt{\left(\lambda+\frac{2}{9\lambda}\right)^2 + \left(\frac{2}{3}\right)^2}} dr$$

(2.11)

$$= \frac{1}{2}\rho U^2 C \left[\frac{2}{3} C_L - \left(\lambda+\frac{2}{9\lambda}\right) C_D \right] \sqrt{\left(\lambda+\frac{2}{9\lambda}\right)^2 + \left(\frac{2}{3}\right)^2} \, dr$$

The total torque of blade element may be derived with Formula (2.11).

$$dM = rdF = \frac{1}{2}\rho U^2 C\left[\frac{2}{3}C_L - \left(\lambda + \frac{2}{9\lambda}\right)C_D\right]\sqrt{\left(\lambda + \frac{2}{9\lambda}\right)^2 + \left(\frac{2}{3}\right)^2} \, rdr \quad (2.12)$$

The total power of blade element may be derived from Formula (2.12) and the relational expression $\omega r = \lambda U$.

$$dP = \omega dM = \frac{1}{2}\rho U^3 C\lambda\left[\frac{2}{3}C_L - \left(\lambda + \frac{2}{9\lambda}\right)C_D\right]\sqrt{\left(\lambda + \frac{2}{9\lambda}\right)^2 + \left(\frac{2}{3}\right)^2} \, dr \quad (2.13)$$

It is needed to substitute the specific formulas of chord C, lift coefficient C_L and drag coefficient C_D into the above formulas and convert the local linear velocity ration λ therein into tip speed ratio λ_t with the conversion formula $\lambda = \lambda_t r/R = \lambda_t x$, and then do integration calculations.

2.4 Integral formula of the pneumatic performance of wind turbine

A wind turbine is composed of several blades, and the performance of wind turbine can be derived by integrating the above blade element performance formulas. The four most important parameters of horizontal axis wind turbine are power coefficient, torque coefficient, lift coefficient and thrust coefficient. The performance of wind turbine which consists of B blades is to be derived below and x represents relative chord r/R in derivation process.

The integral formula of the thrust coefficient of wind turbine is derived from Formula (2.10).

$$\begin{aligned}
C_T &= \frac{B}{\frac{1}{2}\rho U^2 \pi R^2} \int_R \frac{1}{2}\rho U^2 C\left[\left(\lambda + \frac{2}{9\lambda}\right)C_L + \frac{2}{3}C_D\right]\sqrt{\left(\lambda + \frac{2}{9\lambda}\right)^2 + \left(\frac{2}{3}\right)^2} \, dr \\
&= \frac{B}{\pi}\int_0^1 \left(\frac{C}{R}\right)\left[\left(\lambda + \frac{2}{9\lambda}\right)C_L + \frac{2}{3}C_D\right]\sqrt{\left(\lambda + \frac{2}{9\lambda}\right)^2 + \left(\frac{2}{3}\right)^2} \, d\left(\frac{r}{R}\right) \quad (2.14) \\
&= \frac{B}{\pi}\int_0^1 \left(\frac{C}{R}\right)\left[\left(\lambda_t x + \frac{2}{9\lambda_t x}\right)C_L + \frac{2}{3}C_D\right]\sqrt{\left(\lambda + \frac{2}{9\lambda_t x}\right)^2 + \left(\frac{2}{3}\right)^2} \, dx
\end{aligned}$$

The integral formula of the lift coefficient of wind turbine is derived from Formula (2.11).

$$C_F = \frac{B}{\frac{1}{2}\rho U^2 \pi R^2} \int_R \frac{1}{2}\rho U^2 C \left[\frac{2}{3}C_L - \left(\lambda + \frac{2}{9\lambda}\right)C_D\right] \sqrt{\left(\lambda + \frac{2}{9\lambda}\right)^2 + \left(\frac{2}{3}\right)^2} \, dr$$

$$= \frac{B}{\pi} \int_0^1 \left(\frac{C}{R}\right) \left[\frac{2}{3}C_L - \left(\lambda + \frac{2}{9\lambda}\right)C_D\right] \sqrt{\left(\lambda + \frac{2}{9\lambda}\right)^2 + \left(\frac{2}{3}\right)^2} \, d\left(\frac{r}{R}\right) \quad (2.15)$$

$$= \frac{B}{\pi} \int_0^1 \left(\frac{C}{R}\right) \left[\frac{2}{3}C_L - \left(\lambda_1 x + \frac{2}{9\lambda_1 x}\right)C_D\right] \sqrt{\left(\lambda_1 x + \frac{2}{9\lambda_1 x}\right)^2 + \left(\frac{2}{3}\right)^2} \, dx$$

The integral formula of the torque coefficient of wind turbine is derived from Formula (2.12).

$$C_M = \frac{B}{\frac{1}{2}\rho U^2 \pi R^3} \int_R \frac{1}{2}\rho U^2 C \left[\frac{2}{3}C_L - \left(\lambda + \frac{2}{9\lambda}\right)C_D\right] \sqrt{\left(\lambda + \frac{2}{9\lambda}\right)^2 + \left(\frac{2}{3}\right)^2} \, r \, dr$$

$$= \frac{B}{\pi} \int_0^1 \left(\frac{r}{R}\right)\left(\frac{C}{R}\right) \left[\frac{2}{3}C_L - \left(\lambda + \frac{2}{9\lambda}\right)C_D\right] \sqrt{\left(\lambda + \frac{2}{9\lambda}\right)^2 + \left(\frac{2}{3}\right)^2} \, d\left(\frac{r}{R}\right) \quad (2.16)$$

$$= \frac{B}{\pi} \int_0^1 x \left(\frac{C}{R}\right) \left[\frac{2}{3}C_L - \left(\lambda_1 x + \frac{2}{9\lambda_1 x}\right)C_D\right] \sqrt{\left(\lambda_1 x + \frac{2}{9\lambda_1 x}\right)^2 + \left(\frac{2}{3}\right)^2} \, dx$$

The integral formula of the power coefficient of wind turbine is derived from Formula (2.13).

$$C_P = \frac{B}{\frac{1}{2}\rho U^3 \pi R^2} \int_R \frac{1}{2}\rho U^3 \lambda C \left[\frac{2}{3}C_L - \left(\lambda + \frac{2}{9\lambda}\right)C_D\right] \sqrt{\left(\lambda + \frac{2}{9\lambda}\right)^2 + \left(\frac{2}{3}\right)^2} \, dr$$

$$= \frac{B}{\pi} \int_0^1 \lambda \left(\frac{C}{R}\right) \left[\frac{2}{3}C_L - \left(\lambda + \frac{2}{9\lambda}\right)C_D\right] \sqrt{\left(\lambda + \frac{2}{9\lambda}\right)^2 + \left(\frac{2}{3}\right)^2} \, d\left(\frac{r}{R}\right) \quad (2.17)$$

$$= \frac{B}{\pi} \int_0^1 \lambda_1 x \left(\frac{C}{R}\right) \left[\frac{2}{3}C_L - \left(\lambda_1 x + \frac{2}{9\lambda_1 x}\right)C_D\right] \sqrt{\left(\lambda_1 x + \frac{2}{9\lambda_1 x}\right)^2 + \left(\frac{2}{3}\right)^2} \, dx$$

These formulas apply to horizontal axis wind turbines with any airfoil. It is needed to substitute the specific formulas of chord C/R, airfoil lift coefficient C_L and drag coefficient C_D into the above formulas and then do integration calculations. In this way, the optimal pneumatic performance of wind turbine can be derived with analytic calculation if the expressions of chord and angle of attack are known.

2.5 Summary of this chapter

In this chapter, the stable operation conditions of wind turbine are defined as design conditions. These are very important and special operating conditions which are deemed as the target optimal operating conditions of the designed wind turbine, and in such conditions the axial speed induction factor is a constant and thus eliminates the necessity of iterative calculation, which means greatly simplified calculation steps.

In this chapter, the force on the blade of wind turbine in design conditions is analyzed to obtain the basic relational expressions among a number of parameters, such as, among the inflow angle, twist and angle of attack, and among the inflow velocity, inflow angle and tip speed ratio. These relational expressions enable derivation of the differential formula of the pneumatic performance of blade element and the integral formula of the pneumatic performance of wind turbine as per the BEM Theory. The final results of the power, torque, lift and thrust coefficients of wind turbine can be obtained by substituting the specific expressions of the chord formula, airfoil lift coefficient and drag coefficient that are to be derived in the following chapters into the formulas in this chapter and doing integral calculations. Therefore, the contents in this chapter exist as the theoretical basis of what is explored in the following chapters.

Chapter 3   Structure of the Blade of Ideal Wind Turbine

Human beings are eager to understand the structure and performance of ideal wind turbine as a pursuit in order to make better wind turbines. This chapter is intended to put forward the concepts of ideal blade and ideal wind turbine, explore the structural shape of ideal blade, and focus on the research on the distribution function of ideal chord and ideal twist. The performance of ideal wind turbine that consists of ideal blades is to be explored in next chapter.

3.1   The meaning of ideal blade

The ideal blade and its optimal angle of attack, ideal twist and ideal chord will be discussed in this chapter, and now the relevant basic concepts are defined as follows.

Optimal angle of attack: the angle of attack that gives rise to the highest efficiency (namely the maximum power) of blade element is the optimal angle of attack. It will be proven in Section 3.2 that the optimal angle of attack is also the angle that promises the largest airfoil lift drag ratio.

Ideal twist: assuming the airfoil does not change in spanwise direction, the spanwise distribution of the difference between the inflow angle and the optimal angle of blade in design conditions is called ideal twist.

Ideal chord: assuming the airfoil does not change in spanwise direction and the twist is ideal twist, the spanwise distribution of the chord of blade derived as per BEM Theory[24] in design conditions is ideal chord.

Ideal blade: the structure of blade includes twist, chord and airfoil. The blade of which the airfoil structure has ideal twist and ideal chord and has infinite lift drag ratio in ideal fluid is called ideal blade. As ideal fluid does not generate drag, and the lift drag ratio of any airfoil is infinite, the blade which has ideal twist, ideal chord and actual airfoil structure may be deemed as ideal blade.

Ideal wind turbine: the rotor that consists of infinite number of ideal blades is called ideal wind turbine.

3.2 Calculating the optimal angle of attack

Once the airfoil of wind turbine is selected, the lift coefficient C_L, drag coefficient C_D and their ratio ζ (known as lift drag ratio) only change with the change of angle of attack α. Obviously, there is an angle of attack that gives rise to the highest blade efficiency, namely the optimal angle of attack.

Formulas (2.11) and (2.10) provide the total circumferential lift dF and total axial thrust dT of blade element in stable operating conditions:

$$dF = dL_w - dD_w = dL\sin\varphi - dD\cos\varphi = \frac{1}{2}\rho v^2 C(C_L \sin\varphi - C_D \cos\varphi) dr$$

$$= \frac{1}{2}\rho U^2 C\left[\frac{2}{3}C_L - \left(\lambda + \frac{2}{9\lambda}\right)C_D\right]\sqrt{\left(\lambda + \frac{2}{9\lambda}\right)^2 + \left(\frac{2}{3}\right)^2}\, dr$$

(3.1)

$$dT = dL_u + dD_u = dL\cos\varphi + dD\sin\varphi = \frac{1}{2}\rho v^2 C(C_L \cos\varphi + C_D \sin\varphi) dr$$

$$= \frac{1}{2}\rho U^2 C\left[\left(\lambda + \frac{2}{9\lambda}\right)C_L + \frac{2}{3}C_D\right]\sqrt{\left(\lambda + \frac{2}{9\lambda}\right)^2 + \left(\frac{2}{3}\right)^2}\, dr$$

(3.2)

According to Formulas (3.1) and (3.2), the ratio of total lift and total thrust is as follows for any blade element in stable operation state.

$$\frac{dF}{dT} = \frac{\frac{2}{3}C_L - \left(\lambda + \frac{2}{9\lambda}\right)C_D}{\frac{2}{3}C_D + \left(\lambda + \frac{2}{9\lambda}\right)C_L}$$

(3.3)

The blade element efficiency η of blade is proportional to the ratio dF/dT of total lift and total thrust of blade element[25], that is

$$\eta = \frac{dP}{dP_u} = \frac{\omega r\, dF}{U\, dT} = \lambda \frac{dF}{dT}$$

(3.4)

where dP_u is the power imposed by wind onto blade element dr, and dP is the output power of rotor in the portion dr of blade element. For any blade element, the linear velocity ratio λ does not change with the change of angle of attack in stable operation state. To derive the extreme point, let

$$\frac{\partial \eta}{\partial \alpha} = \lambda \frac{\partial}{\partial \alpha}\left(\frac{dF}{dT}\right) = 0 \tag{3.5}$$

Substitute Formula (3.3) into Formula (3.5) to get

$$\frac{(81\lambda^4 + 72\lambda^2 + 4)[C'_L(\alpha)C_D(\alpha) - C_L(\alpha)C'_D(\alpha)]}{[6\lambda C_D(\alpha) + (9\lambda^2 + 2)C_L(\alpha)]^2} = 0 \tag{3.6}$$

It is derived from this formula:

$$\frac{C_L(\alpha)}{C_D(\alpha)} = \frac{C'_L(\alpha)}{C'_D(\alpha)} \tag{3.7}$$

or

$$\frac{C_L(\alpha)}{C_D(\alpha)} = \frac{dC_L(\alpha)}{dC_D(\alpha)} \tag{3.8}$$

The angle of attack that fulfills Formulas (3.7) or (3.8) is the optimal angle of attack. This angle of attack is the angle of attack at tangency point where the Eiffel Polar Curve[26] (the airfoil lift and drag coefficient relationship curve) is tangent to the tangent line crossing the origin point, and expressed as α_b (Figure 3.1).

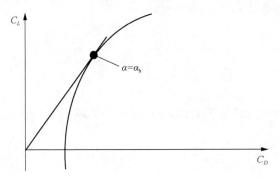

Figure 3.1 Schematic diagram of the position of optimal angle of attack of Eiffel Polar Curve

It is shown in Figure 3.1 that, for actual airfoil, there is only one optimal angle of attack, and it is proven below that the optimal angle of attack is the one that results in the maximum lift drag ratio. The lift drag ratio ζ of blade element is defined as

$$\zeta = \frac{C_L(\alpha)}{C_D(\alpha)} \tag{3.9}$$

Supposing $\partial \zeta/\partial \alpha$, it is derived that

$$\frac{C_L(\alpha)}{C_D(\alpha)} = \frac{C'_L(\alpha)}{C'_D(\alpha)} \tag{3.10}$$

Formula (3.12) is the same as Formula (3.7), which proves that the optimal angle of attack is the one that results in maximum blade element lift drag ratio. This conclusion reveals that, to reach the highest efficiency of wind turbine, the angle of attack in each position of blade should lead to the maximum lift drag ratio in this position, namely optimal angle of attack should exist in each position and this can be realized through reasonable design of twist.

3.3 Calculation of ideal twist

For given design tip speed ratio λ_t the ratio $\omega R/U$ of tip linear velocity ωR and the velocity U of the flow from an infinite point of wind turbine, the speed induction factor and inflow angle of any blade element are definite in stable operation state and remain the same for any wind turbine. According to Figure 2.2 and Formula (2.3), the inflow angle φ can be calculated with the formula below[27]:

$$\tan \varphi = \frac{u}{w} = \frac{1-a}{1+\frac{a(1-a)}{\lambda^2}} \frac{1}{\lambda} = \frac{6\lambda}{9\lambda^2 + 2} = \frac{6\lambda_t x}{9\lambda_t^2 x^2 + 2} \tag{3.11}$$

where, $x=r/R$ is relative radius. There is maximum inflow angle when $x = \sqrt{2}/(3\lambda_t)$.

$$\varphi_{max} = \arctan\left(\sqrt{2}/2\right) \approx 35.3° \tag{3.12}$$

The larger the tip speed ratio is, the closer the maximum value point is to the blade root. The maximum value point is on the inner side of $0.1R$ when tip speed ratio exceeds 5. Supposing the given tip speed ratio λ_t is 6, 8 and 10 respectively, the spanwise change of the inflow angle φ along r/R is as shown in Figure 3.2.

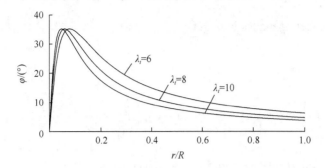

Figure 3.2　The trend of inflow angle curve

Twist and angle of attack are mutually complementary and their sum equals to the inflow angle, so the twist that fulfills the optimal angle of attack α_b is ideal twist. According to Formula (3.11), the calculation formula of ideal twist β is

$$\beta = \varphi - \alpha_b = \arctan\frac{6\lambda_t x}{9\lambda_t^2 x^2 + 2} - \alpha_b \tag{3.13}$$

It is shown in the above formula that the ideal twist is determined by design tip speed ratio λ_t and optimal angle of attack α_b.

For example, if the design adopts NACA 23015 airfoil of which the optimal angle of attack α_b is about 6° without spanwise change, then the expression of ideal twist is

$$\beta = \varphi - \alpha_b = \arctan\frac{6\lambda_t x}{9\lambda_t^2 x^2 + 2} - 6° \tag{3.14}$$

It is visible that the optimal angle of attack can be obtained through inverse derivation with analytic calculation.

3.4　Calculation of ideal chord

According to momentum theory, the thrust imposed by the incoming flow onto the annulus of which the radius is r to $r+\mathrm{d}r$ of the rotor disk is[28,29]

$$\mathrm{d}T = 4\pi\rho U^2 a(1-a)r\,\mathrm{d}r \tag{3.15}$$

Substituting the axial speed induction factor $a=1/3$ in stable operation state and making it equal to the thrust Formula (2.10) derived as per blade element theory, assuming the number of blades is B, get

$$4\pi\rho U^2 \cdot \frac{1}{3}\left(1-\frac{1}{3}\right)r\,dr$$

$$= B \cdot \frac{1}{2}\rho U^2 C\left[\left(\lambda+\frac{2}{9\lambda}\right)C_L + \frac{2}{3}C_D\right]\sqrt{\left(\lambda+\frac{2}{9\lambda}\right)^2 + \left(\frac{2}{3}\right)^2}\,dr \qquad (3.16)$$

The expression of relative chord can be solved hereby

$$\frac{C}{R} = \frac{16\pi}{9B}\frac{r}{R}\frac{1}{\left[\left(\lambda+\frac{2}{9\lambda}\right)C_L + \frac{2}{3}C_D\right]\sqrt{\left(\lambda+\frac{2}{9\lambda}\right)^2 + \left(\frac{2}{3}\right)^2}} \qquad (3.17)$$

The spanwise distribution expression of ideal relative chord can be derived by substituting the lift and drag coefficients of corresponding optimal angle of attack:

$$\frac{C}{R} = \frac{16\pi}{9B}\frac{x}{\left[\left(\lambda_t x+\frac{2}{9\lambda_t x}\right)C_L(\alpha_b) + \frac{2}{3}C_D(\alpha_b)\right]\sqrt{\left(\lambda_t x+\frac{2}{9\lambda_t x}\right)^2 + \left(\frac{2}{3}\right)^2}} \qquad (3.18)$$

It is shown in Formula (3.18) that ideal relative chord is a function of design tip speed ratio and lift and drag coefficients of optimal angle of attack. Formula (3.18) is derived based on BEM Theory and applicable to all airfoils.

Supposing the number of blades $B=3$, if NACA 23015 airfoil is adopted, then the optimal angle of attack α_b of this airfoil is about 6°, the corresponding lift coefficient is $C_L=0.76$ and drag coefficient is $C_D=0.0087$; substitute Formula (3.18) to get specific expression:

$$\frac{C}{R} = \frac{16\pi}{27}\frac{x}{\left[0.76\left(\lambda_t x+\frac{2}{9\lambda_t x}\right) + 0.008,7\times\frac{2}{3}\right]\sqrt{\left(\lambda_t x+\frac{2}{9\lambda_t x}\right)^2 + \left(\frac{2}{3}\right)^2}} \qquad (3.19)$$

For different given design tip speed ratio, the curvilinear curve of ideal relative chord of airfoil NACA 23015 is shown in Figure 3.3.

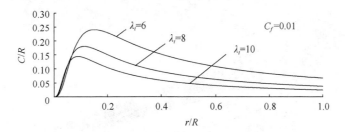

Figure 3.3 Example of curvilinear curve of ideal relative chord

It is visible that the ideal chord can be obtained through inverse derivation with analytic calculation. The figure shows that it is very difficult to process and produce such distorted shape, and considering the curve of twist, it is beyond imagination how difficult it is to manufacture it. But as ideal chord formula, it may guide the design and plays a very important theoretical role. For example, it may be used to derive extreme parameter values such as power, torque, thrust and other performance indicators.

3.5 About ideal airfoil

The ideal structure of airfoil should be the structure that enables infinite lift drag ratio or that has 0 drag. Obviously such structural form does not exist and of course it is hard to express it with graph or function. But the lift drag ratio of actual airfoil is a finite value and its structural forms can be expressed with graphs and some forms can be expressed with functions. The way how to express airfoil with functions will be explored in Chapter 7.

The lift drag ratio of actual airfoil cannot be an infinite value, but the airfoil running in ideal fluid has 0 drag and infinite lift drag ratio; therefore, any actual airfoil running in ideal fluid may be deemed as ideal airfoil.

3.6 The form of ideal blade

The structural form of ideal blade is determined by blade function which consists of three subfunctions: chord function, twist function and airfoil function. An example of an ideal blade structure is shown in Figure 3.4. Its twist function is determined by Formula (12.11), chord function is determined by Formula (12.12) and airfoil function is determined by Formulas (12.9) and (12.10) (Chapter 12).

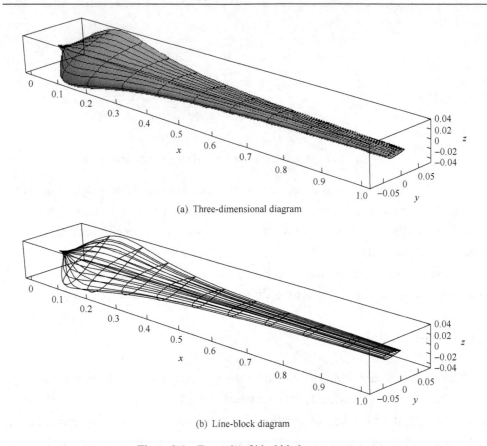

(a) Three-dimensional diagram

(b) Line-block diagram

Figure 3.4 Example of ideal blade structure

Ideal blade is of great theoretical value, but it cannot be used directly, and only after being modified by taking a number of factors such as structural strength and manufacturing/installation technique into consideration will it be of practical value. See Chapter 12 for the example of the structural form of practical blade. This chapter systematically explores the functional design method for blade, namely it studies how to generate three-dimensional graphs with blade functions to draw the pictorial drawing of blade in order to perform blade design.

3.7 Summary of this chapter

This chapter starts with the definition of basic concepts of the optimal angle of attack, ideal twist and ideal chord. It is also proven in this chapter that the angle of attack that promises the highest efficiency of blade element is the angle of attack that ensures the maximum lift drag ratio, and the analytic calculation formulas of ideal twist and ideal chord are also derived with the BEM Theory. The research shows that ideal chord is a function of tip speed ratio, lift and drag coefficient and number of blades, and ideal twist is the difference of inflow angle and the optimal angle of attack. This chapter also puts forward the concepts of ideal blade and ideal blade and ideal wind turbine based on the concepts of ideal twist and ideal chord, and the structural characteristics of ideal blade are clarified, laying the foundation for establishing ideal wind turbine theory.

Chapter 4 Highest Performance of Ideal Wind Turbine

The highest performance of ideal wind turbine refers to the power, torque, lift and thrust performance of wind turbine running in ideal fluid. As ideal fluid does not generate drag (the lift drag ratio of airfoil is infinite), the performance of ideal wind turbine in ideal fluid is merely a function of lift drag ratio. In this case the performance of ideal wind turbine is the highest performance and an insurmountable performance limit, so it is of theoretical significance to explore such highest performance.

According to Betz Limit, the maximum value of the power coefficient of wind turbine is about 0.593, and even so it remains an insurmountable limit. The power coefficient of modern wind turbines has never exceeded 0.5[30,31]. Is there another limit? It is to be proven in this chapter that such limit do exists, and it is related to the design tip speed ratio of wind turbine. Only when the tip speed ratio approaches infinity, will such related limit approaches the Betz Limit. Similarly, the limits of lift, torque and thrust performance of wind turbine will be provided in this chapter.

Now the environmental characteristics based on which the highest performance and general performance of ideal wind turbine and practical wind turbine are calculated and related sections are listed in Table 4.1, so that the readers can have a deep understanding of the differences and relations among chapters (contents). For example, the performance of ideal wind turbine is also explored in Chapter 5, and the difference between it and this chapter lies in that the ideal wind turbine operates in actual fluid environment (lift drag ratio is no longer infinite) in order to study the general performance of ideal wind turbine.

Table 4.1 Environment for calculating different performance of wind turbine and layout of chapters

Types of wind turbine	Highest performance	General performance
Ideal wind turbine	Ideal blade structure (ideal chord and twist) Infinite number of blades (irrespective of tip loss) Ideal liquid environment (lift drag ratio is infinite) (Chapter 4)	Ideal blade structure (ideal chord and twist) Infinite number of blades (irrespective of tip loss) Actual liquid environment (lift drag ratio is infinite) (Chapter 5)

Continued

Types of wind turbine	Highest performance	General performance
Practical wind turbine	Ideal blade structure (ideal chord and twist) Finite number of blades (take tip loss into consideration) Actual liquid environment (lift drag ratio is infinite) (Chapter 9)	Simplified blade structure (simplified chord and twist) Finite number of blades (take tip loss into consideration) Actual liquid environment (lift drag ratio is infinite) (Chapter 11)

4.1 Power performance and its limit

Supposing wind turbine is composed of B ideal blades, substitute the angle of attack of ideal blade α_b and chord Formula (3.18) into the analytic expression of blade element power in Formula (2.17) and execute integration to derive the power coefficient calculation expression of wind turbine[32].

$$C_P = \frac{B}{\frac{1}{2}\rho U^3 \pi R^2} \int_R dP = \frac{B}{\pi} \int_R \left(\frac{C}{R}\right) \lambda \left[\frac{2}{3}C_L - \left(\lambda + \frac{2}{9\lambda}\right)C_D\right] \sqrt{\left(\lambda + \frac{2}{9\lambda}\right)^2 + \left(\frac{2}{3}\right)^2} d\left(\frac{r}{R}\right)$$

$$= \frac{B}{\pi} \int_0^1 \left\{ \frac{16\pi}{9B} \cdot \frac{x}{\left[\left(\lambda_t x + \frac{2}{9\lambda_t x}\right)C_L(\alpha_b) + \frac{2}{3}C_D(\alpha_b)\right]\sqrt{\left(\lambda_t x + \frac{2}{9\lambda_t x}\right)^2 + \left(\frac{2}{3}\right)^2}} \right\} $$

$$\cdot \lambda_t x \left[\frac{2}{3}C_L(\alpha_b) - \left(\lambda_t x + \frac{2}{9\lambda_t x}\right)C_D(\alpha_b)\right] \sqrt{\left(\lambda_t x + \frac{2}{9\lambda_t x}\right)^2 + \left(\frac{2}{3}\right)^2} dx$$

$$= \frac{16}{9} \int_0^1 \frac{\lambda_t x^2 \left[\frac{2}{3}C_L(\alpha_b) - \left(\lambda_t x + \frac{2}{9\lambda_t x}\right)C_D(\alpha_b)\right]}{\left(\lambda_t x + \frac{2}{9\lambda_t x}\right)C_L(\alpha_b) + \frac{2}{3}C_D(\alpha_b)} dx$$

(4.1)

The integral result of this expression is too complex, so it is optional to substitute the corresponding specific values of the lift coefficient C_L and drag coefficient C_D of optimal angle of attack α_b, and then execute integration to get the formula of power coefficient in design situation. If expressed with symbol, the integral result of Formula (4.1) is

$$C_P = \left[\frac{64C_D\left(-2C_D^4 + C_L^2C_D^2 + 3C_L^4\right)}{243\lambda_t^2 C_L^4 \sqrt{2C_L^2 - C_D^2}} \arctan\frac{C_D + 3C_L\lambda_t x}{\sqrt{2C_L^2 - C_D^2}}\right]_{x=0}^1$$

$$+ \left[\frac{32\left(2C_D^4 + C_L^2C_D^2 - C_L^4\right)}{243\lambda_t^2 C_L^4} \ln\left(2C_L + 6C_D\lambda_t x + 9C_L\lambda_t^2 x^2\right)\right]_{x=0}^1$$

$$+ \left[\frac{16}{81} \frac{\left(C_D^2 + C_L^2\right)\left(3C_L\lambda_t x^2 - 4C_D x\right) - 3C_L^2 C_D\lambda_t^2}{\lambda_t C_L^3}\right]_{x=0}^1 \quad (4.2)$$

$$= \frac{64C_D\left(2C_D^4 - C_L^2C_D^2 - 3C_L^4\right)}{243\lambda_t^2 C_L^4 \sqrt{2C_L^2 - C_D^2}} \left[\arctan\frac{C_D}{\sqrt{2C_L^2 - C_D^2}} - \arctan\frac{C_D + 3\lambda_t C_L}{\sqrt{2C_L^2 - C_D^2}}\right]$$

$$+ \frac{32\left(2C_D^4 + C_L^2C_D^2 - C_L^4\right)}{243\lambda_t^2 C_L^4} \ln\frac{2C_L + 6\lambda_t C_D + 9\lambda_t^2 C_L}{2C_L}$$

$$+ \frac{16\left[3\lambda_t C_L C_D^2 - 4C_D^3 + 3\lambda_t C_L^3 - \left(3\lambda_t^2 + 4\right)C_L^2 C_D\right]}{81\lambda_t C_L^3}$$

This is the power coefficient formula of ideal wind turbine. Now let's take a look at the characteristics of power coefficient with 0 drag. In Formulas (4.1) or (4.2), drag coefficient $C_D = 0$ is the special case (i.e., the lift drag ratio is infinite) in which the power coefficient of ideal wind turbine reaches up to the maximum value with changing design tip speed ratio. Supposing $C_D = 0$, integrate Formula (4.1) to get

$$C_P\big|_{C_D=0} = \frac{16}{9}\int_0^1 \frac{\frac{2}{3}\lambda_t x^2}{\lambda_t x + \frac{2}{9\lambda_t x}} dx = \frac{16}{243\lambda_t^2}\left[9\lambda_t^2 - 2\ln\left(9\lambda_t^2 + 2\right) + 2\ln 2\right] \quad (4.3)$$

In this book Formula (4.3) is called the formula of maximum power coefficient of ideal wind turbine related to design tip speed ratio. Figure 4.1 reflects the changing trend of the limit related to tip speed ratio. Being different from Betz Limit which is a horizontal straight line, the limit related to tip speed ratio is a curve and it is a little bit smaller than the fixed value by Betz Limit, so the limit related to tip speed ratio is of more practical value than Betz Limit.

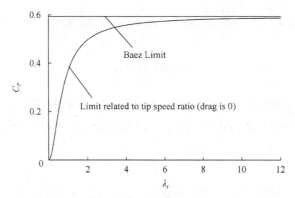

Figure 4.1 The changing trend of the formula of maximum power coefficient of ideal wind turbine related to design tip speed ratio

According to Formula (4.3), this formula approaches the Betz Limit when the tip speed ratio increases infinitely:

$$C_{P\max}\big|_{C_D=0} = \lim_{\lambda_t \to \infty} C_P\big|_{C_D=0} = \lim_{\lambda_t \to \infty} \frac{16\partial_{\lambda_t}\left[9\lambda_t^2 - 2\ln\left(2+9\lambda_t^2\right)+\ln 4\right]}{243\partial_{\lambda_t}\left(\lambda_t^2\right)}$$

$$= \lim_{\lambda_t \to \infty} \frac{16\left(18\lambda_t - \dfrac{36\lambda_t}{2+9\lambda_t^2}\right)}{486\lambda_t} = \frac{16}{27}$$

(4.4)

It is visible that only when both lift drag ratio and tip speed ratio approach infinite value will the power coefficient approaches Betz Limit.

The method used to put forward Betz Limit is Momentum Theory, and the last step of derivation is described as follows[33]: the rotor is deemed as a disk and the upstream/downstream differential pressure is multiplied by the velocity of incoming flow to obtain dimensionless power coefficient formula.

$$C_P = 4a(1-a)^2 \tag{4.5}$$

When extreme value of this formula is evaluated to get the axial induction speed $a = 1/3$, the maximum value of power coefficient is 16/27 which is the Betz Limit. This derivation process involves neither the shape of blade or other details nor the relation between power coefficient and tip speed ratio and lift drag ratio, and what is obtained is the final value in the most ideal state (both lift drag ratio and tip speed ratio approach infinite value). Betz Limit is demonstrated and elaborated with analytic calculation method herein, and this process involves the derivation and calculation of almost all important parameters such as angle of attack and chord. Seeing from another perspective, Betz Limit is a special case of formula in this book only.

4.2 Torque performance and its limit

Substitute Formula (3.18) of ideal chord into Formula (2.16) of wind turbine torque coefficient and formula to derive the formula of ideal wind turbine torque coefficient[34].

$$C_M = \frac{B}{\pi} \int_0^1 x \left(\frac{C}{R}\right) \left[\frac{2}{3} C_L - \left(\lambda_t x + \frac{2}{9\lambda_t x}\right) C_D\right] \sqrt{\left(\lambda_t x + \frac{2}{9\lambda_t x}\right)^2 + \left(\frac{2}{3}\right)^2} \, dx$$

$$= \frac{B}{\pi} \int_0^1 x \left\{ \frac{\dfrac{16\pi}{9B}}{\left[\left(\lambda_t x + \dfrac{2}{9\lambda_t x}\right) C_L(\alpha_b) + \dfrac{2}{3} C_D(\alpha_b)\right] \sqrt{\left(\lambda_t x + \dfrac{2}{9\lambda_t x}\right)^2 + \left(\dfrac{2}{3}\right)^2}} \right\}$$

$$\left[\frac{2}{3} C_L(\alpha_b) - \left(\lambda_t x + \frac{2}{9\lambda_t x}\right) C_D(\alpha_b)\right] \sqrt{\left(\lambda_t x + \frac{2}{9\lambda_t x}\right)^2 + \left(\frac{2}{3}\right)^2} \, dx \quad (4.6)$$

$$= \frac{16}{9} \int_0^1 \frac{x^2 \left[\dfrac{2}{3} C_L(\alpha_b) - \left(\lambda_t x + \dfrac{2}{9\lambda_t x}\right) C_D(\alpha_b)\right]}{\left(\lambda_t x + \dfrac{2}{9\lambda_t x}\right) C_L(\alpha_b) + \dfrac{2}{3} C_D(\alpha_b)} \, dx$$

Drag coefficient $C_D = 0$ is the special case in which the torque coefficient of ideal wind turbine reaches up to the maximum value with changing design tip speed ratio. Supposing $C_D=0$, integrate the above formula

$$C_{M\,max} = \frac{16}{9} \int_0^1 \frac{\dfrac{2}{3} x^2}{\left(\lambda_t x + \dfrac{2}{9\lambda_t x}\right)} \, dx = \frac{16}{243\lambda_t^3} \left[9\lambda_t^2 - 2\ln\left(9\lambda_t^2 + 2\right) + 2\ln 2\right] \quad (4.7)$$

The highest performance formula for the torque coefficient related to design tip speed ratio when drag is 0 is obtained hereby. Theoretically speaking, for the same design tip speed ratio, the torque coefficient of any wind turbine will not exceed the top limit specified by this formula in stable operation state, regardless of how much the drag is reduced.

The curve of theoretical limit of torque coefficient that corresponds to the design tip speed ratio of wind turbine is shown in Figure 4.2.

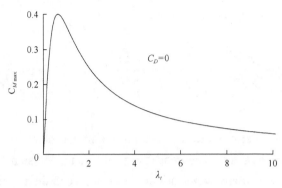

Figure 4.2   Theoretical limit of torque coefficient related to tip speed ratio

When design tip speed ratio is about 0.635,428, this curve reaches up to its maximum value which is about 0.401,017.

4.3   Lift performance and its limit

Substitute Formula (3.18) of ideal chord into Formula (2.15) of wind turbine lift coefficient and execute integration to derive the formula of ideal wind turbine lift coefficient[35].

$$C_F = \frac{B}{\pi} \int_0^1 \left(\frac{C}{R}\right) \left[\frac{2}{3} C_L - \left(\lambda_t x + \frac{2}{9\lambda_t x}\right) C_D \right] \sqrt{\left(\lambda_t x + \frac{2}{9\lambda_t x}\right)^2 + \left(\frac{2}{3}\right)^2}\, dx$$

$$= \frac{B}{\pi} \int_0^1 \left\{ \frac{\frac{16\pi}{9B}}{\left[\left(\lambda_t x + \frac{2}{9\lambda_t x}\right) C_L(\alpha_b) + \frac{2}{3} C_D(\alpha_b)\right] \sqrt{\left(\lambda_t x + \frac{2}{9\lambda_t x}\right)^2 + \left(\frac{2}{3}\right)^2}} \right. \times$$

$$\left. \left[\frac{2}{3} C_L(\alpha_b) - \left(\lambda_t x + \frac{2}{9\lambda_t x}\right) C_D(\alpha_b)\right] \sqrt{\left(\lambda_t x + \frac{2}{9\lambda_t x}\right)^2 + \left(\frac{2}{3}\right)^2}\, dx \right\} \quad (4.8)$$

$$= \frac{16}{9} \int_0^1 \frac{x \left[\frac{2}{3} C_L(\alpha_b) - \left(\lambda_t x + \frac{2}{9\lambda_t x}\right) C_D(\alpha_b)\right]}{\left(\lambda_t x + \frac{2}{9\lambda_t x}\right) C_L(\alpha_b) + \frac{2}{3} C_D(\alpha_b)}\, dx$$

Drag coefficient $C_D = 0$ is the special case in which the lift coefficient of ideal wind turbine reaches up to the maximum value with changing design tip speed ratio. Supposing $C_D=0$, integrate the above formula

$$C_{F\max} = \frac{16}{9}\int_0^1 \frac{\frac{2}{3}x}{\lambda_t x + \frac{2}{9\lambda_t x}} dx = \frac{32}{81\lambda_t^2}\left(3\lambda_t - \sqrt{2}\arctan\frac{3\lambda_t}{\sqrt{2}}\right) \quad (4.9)$$

This is the highest performance formula for the lift coefficient related to design tip speed ratio when the drag is 0. Theoretically speaking, for the same design tip speed ratio, the lift coefficient of any wind turbine will not exceed the top limit specified by this formula in stable operation state, regardless of how much the drag is reduced.

The curve of theoretical limit of lift coefficient that corresponds to the design tip speed ratio of wind turbine is shown in Figure 4.3.

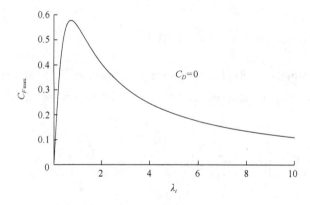

Figure 4.3 Theoretical limit of lift coefficient related to tip speed ratio

When design tip speed ratio is about 0.714,175, this curve reaches up to its maximum value which is about 0.577,950.

4.4 Thrust performance and its limit

Substitute Formula (3.18) of ideal chord into Formula (2.14) of wind turbine thrust coefficient and execute integration to derive the formula of ideal wind turbine axial total thrust.

$$C_T = C_{T\min} = \frac{B}{\pi}\int_0^1 \left(\frac{C}{R}\right)\left[\left(\lambda_1 x + \frac{2}{9\lambda_1 x}\right)C_L + \frac{2}{3}C_D\right]\sqrt{\left(\lambda + \frac{2}{9\lambda_1 x}\right)^2 + \left(\frac{2}{3}\right)^2}\,\mathrm{d}x$$

$$= \frac{B}{\pi}\int_0^1 \left\{\frac{\dfrac{16\pi}{9B}\cdot x}{\left[\left(\lambda_1 x + \dfrac{2}{9\lambda_1 x}\right)C_L(\alpha_b) + \dfrac{2}{3}C_D(\alpha_b)\right]\sqrt{\left(\lambda_1 x + \dfrac{2}{9\lambda_1 x}\right)^2 + \left(\dfrac{2}{3}\right)^2}}\right\} \quad (4.10)$$

$$\left[\left(\lambda_1 x + \frac{2}{9\lambda_1 x}\right)C_L(\alpha_b) + \frac{2}{3}C_D(\alpha_b)\right]\sqrt{\left(\lambda + \frac{2}{9\lambda_1 x}\right)^2 + \left(\frac{2}{3}\right)^2}\,\mathrm{d}x$$

$$= \frac{16}{9}\int_0^1 x\,\mathrm{d}x = \frac{8}{9}$$

It is visible that thrust is irrelevant to chord, tip speed ratio and lift drag ratio in stable operating conditions. It should be noticed that, when the tip speed ratio and lift drag ratio change, the chord derived based on Blade BEM Theory changes accordingly. In the above integrated function, chord and other items are inversely proportional simultaneously and the effect of tip speed ratio and lift drag ratio is eliminated in the result.

The smaller the thrust is, the smaller the force on tower is, the smaller the bending moment at foundation of tower is and the stronger the anti-typhoon capability of wind will be. But small thrust coefficient does not indicate small thrust because thrust is highly related to wind velocity.

4.5 Summary of this chapter

In this chapter, the chord formula of ideal blade is substituted into blade element performance analytic expression and integrated to derive the calculation formulas of power, torque, lift and thrust coefficients; for the blades operating in ideal fluid environment, it is supposed the lift drag ratio approaches an infinite value (drag is 0) to obtain the highest performance formulas of power, torque, lift and thrust, and further it is supposed the tip speed ratio approaches an infinite value to obtain the limit performance formulas of power, torque, lift and thrust.

The highest performance and limit performance of ideal wind turbine in stable operating conditions are summarized in Table 4.2.

Table 4.2 The highest performance and limit performance of ideal wind turbine in stable operating conditions

Performance parameter	Highest performance		Limit performance	
	Expression	Preconditions for obtainment	Limit value	Preconditions for obtainment
Power Coefficient	$\dfrac{16}{243\lambda_t^2}\left[9\lambda_t^2 - 2\ln\left(9\lambda_t^2 + 2\right) + 2\ln 2\right]$	$C_D = 0$	0.592,593	$C_D = 0$ and $\lambda_t \to \infty$
Torque Coefficient	$\dfrac{16}{243\lambda_t^3}\left[9\lambda_t^2 - 2\ln\left(9\lambda_t^2 + 2\right) + 2\ln 2\right]$	$C_D = 0$	0.401,017	$C_D = 0$ and $\lambda_t \approx 0.635,428$
Lift Coefficient	$\dfrac{32}{81\lambda_t^2}\left(3\lambda_t - \sqrt{2}\arctan\dfrac{3\lambda_t}{\sqrt{2}}\right)$	$C_D = 0$	0.577,950	$C_D = 0$ and $\lambda_t \approx 0.714,175$
Thrust Coefficient	8/9 (any design tip speed ratio in stable working conditions)	Any drag	0.888,889	Stable operating conditions

Chapter 5 General Performance of Ideal Wind Turbine

The general performance of ideal wind turbine refers to the power, torque, lift and thrust performance of wind turbine running in actual fluid. Being different from the operation in ideal fluid (Chapter 4), the lift drag ratio of airfoil in actual fluid is a limited value, so the performance of this ideal wind turbine is not only a function of tip speed ratio but also a function of lift drag ratio.

In this case the performance will certainly reduce, but it is more approximate to actual situations. Neither the lift drag ratio nor the tip speed ratio of actual wind turbine approaches infinite value, so the general performance of the wind turbine of which both lift drag ratio and tip speed ratio are limited values is of more practical value, and compared with the highest performance and limit performance of ideal wind turbine, it is more suitable to be the target of designed actual wind turbine.

This chapter is intended to derive the power, torque, lift and thrust performance of ideal wind turbine in the conditions of any tip speed ratio and any lift drag ratio.

5.1 General performance of power

The lift drag ratio of airfoil is

$$\zeta = C_L / C_D \tag{5.1}$$

Formula (5.1) may also be expressed as follows:

$$C_L = \zeta C_D \tag{5.2}$$

Substitute Formula (5.2) into power coefficient Formula (4.1) and make integration to get

$$C_{P\max} = \frac{16}{9} \int_0^1 \frac{\lambda_t x^2 \left[\frac{2}{3}\varsigma - \left(\lambda_t x + \frac{2}{9\lambda_t x}\right)\right]}{\left(\lambda_t x + \frac{2}{9\lambda_t x}\right)\varsigma + \frac{2}{3}} dx$$

$$= \frac{64(2-\varsigma^2-3\varsigma^4)}{243\lambda_t^2 \varsigma^4 \sqrt{2\varsigma^2-1}} \left[\arctan\frac{1}{\sqrt{2\varsigma^2-1}} - \arctan\frac{1+3\lambda_t\varsigma}{\sqrt{2\varsigma^2-1}}\right]$$

$$+ \frac{32(2+\varsigma^2-\varsigma^4)}{243\lambda_t^2 \varsigma^4} \ln\frac{2\varsigma+6\lambda_t+9\lambda_t^2\varsigma}{2\varsigma} + \frac{16\left[3\lambda_t\varsigma-4+3\lambda_t\varsigma^3-\left(3\lambda_t^2+4\right)\varsigma^2\right]}{81\lambda_t\varsigma^3}$$

(5.3)

This is the power coefficient formula of ideal wind turbine operating in actual fluid.

It is shown in Formula (5.3) that neither lift coefficient and drag coefficient appears separately nor both of them are included in lift drag ratio, which indicates that the power performance of wind turbine is irrelevant to the absolute value of the lift and drag coefficients of airfoil but closely related to their ratio. This conclusion is of great significance for design of airfoil, namely the design of airfoil is not solely intended to pursue the maximum lift or minimum lift but to pursue the maximum lift drag ratio. Griffiths pointed out that the lift drag ratio of airfoil has great impacts on power coefficient[36], and it is also pointed out in a literature that each section airfoil must have the maximum power coefficient[37]. These qualitative conclusions are consistent to the quantitative results derived in this book. But it is shown in Formula (8.9) that the higher the lift is, the smaller the chord will be, which is favorable for production cost and safety.

Formula (5.3) may be called the power coefficient formula related to design tip speed ratio and lift drag ratio. As this formula is only related to two important parameters and less than Betz Limit, it is of more practical value. It indicates the top limit reference of power coefficient that can be reached by any designed wind turbine with given tip speed ratio and lift drag ratio.

As Formula (5.3) is rather complex, atlas and charts can be made with this formula to facilitate its use. See Figures 5.1 and 5.2 for the atlas that shows how C_P changes with tip speed ratio λ_t and how C_P changes with lift drag ratio ς.

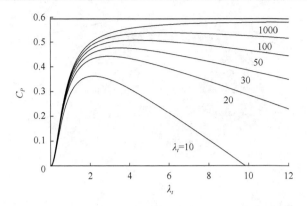

Figure 5.1 Atlas showing how the power coefficient of ideal wind turbine changes with the change of design tip speed ratio

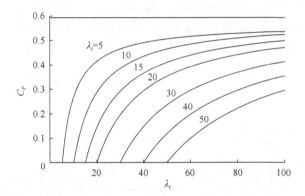

Figure 5.2 Atlas showing how the power coefficient of ideal wind turbine changes with the change of lift drag ratio

The horizontal line in Figures 5.1 and 5.2 is Betz Limit (0.593). This is the curve atlas of ideal wind turbine calculated on the basis of stable operation state (speed induction factor is a stable value), and the reference value of power coefficient that can be reached by wind turbine can be referred as long as lift drag ratio ζ and design tip speed ratio λ_t is given. See Figure 5.3 for the three-dimensional graph that shows how the power coefficient changes with the change of design tip speed ratio λ_t and airfoil lift drag ratio ζ.

It is also shown in Figures 5.1-5.3 that, if the tip speed ratio is determined, the power coefficient will increase as the lift drag ratio of airfoil increases. Obviously, the increase of lift drag ratio of airfoil is of great significance for increase of power coefficient.

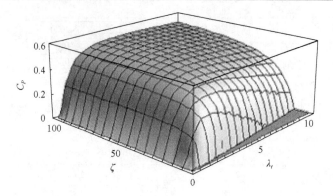

Figure 5.3 Three-dimensional graph showing how the power coefficient of ideal wind turbine changes with change of lift drag ratio and tip speed ratio

Figure 5.4 shows the curve of design tip speed ratio of corresponding maximum power coefficient with given lift drag ratio, and the power coefficient of corresponding point on this curve can be found in Table 5.1.

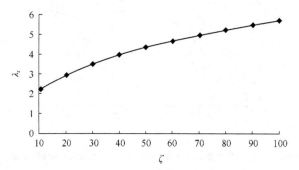

Figure 5.4 The design tip speed ratio-lift drag ratio curve that corresponds to maximum power coefficient

If the lift drag ratio of airfoil is already determined, the tip speed ratio may neither be too high or too low, or it will impair the power coefficient. It is also shown in Figure 5.4 that, the tip speed ratio which leads to the maximum power coefficient of the wind turbine that has a lift drag ratio of less than 100 is not higher than 6, but the tip speed ratio of actual wind turbine will exceed 6 sometimes. Possibly the purpose is to reduce the weight and cost of blades and gear box while compromise a little power coefficient. Such compromise is very common in actual design, but only with theoretical guidance will be such compromise not so aimless.

Table 5.1 The theoretical value of power coefficient formula of ideal wind turbine operating in actual fluid

Lift drag ratio ζ	Design tip speed ratio λ_t										Maximum power coefficient $C_{P\max}$
	1	2	3	4	5	6	7	8	9	10	
∞	0.368	0.496	0.538	0.557	0.568	0.574	0.578	0.581	0.583	0.585	0.593 ($\lambda_t \to \infty$)
1000	0.367	0.494	0.536	0.555	0.565	0.570	0.574	0.576	0.578	0.579	0.580 ($\lambda_t \approx 13.7$)
900	0.367	0.494	0.536	0.554	0.564	0.570	0.573	0.576	0.577	0.578	0.579 ($\lambda_t \approx 13.2$)
800	0.367	0.494	0.536	0.554	0.564	0.569	0.573	0.575	0.576	0.577	0.578 ($\lambda_t \approx 12.6$)
700	0.367	0.494	0.535	0.554	0.563	0.569	0.572	0.574	0.575	0.576	0.576 ($\lambda_t \approx 12.0$)
600	0.367	0.493	0.535	0.553	0.563	0.568	0.571	0.573	0.574	0.575	0.575 ($\lambda_t \approx 11.3$)
500	0.366	0.493	0.534	0.552	0.561	0.567	0.570	0.571	0.572	0.573	0.573 ($\lambda_t \approx 10.5$)
400	0.366	0.492	0.533	0.551	0.560	0.565	0.567	0.569	0.569	0.570	0.570 ($\lambda_t \approx 9.7$)
300	0.365	0.491	0.532	0.549	0.557	0.562	0.564	0.565	0.565	0.565	0.565 ($\lambda_t \approx 8.7$)
200	0.364	0.489	0.528	0.545	0.552	0.556	0.557	0.557	0.556	0.555	0.557 ($\lambda_t \approx 7.4$)
100	0.360	0.482	0.519	0.532	0.537	0.537	0.536	0.533	0.529	0.525	0.538 ($\lambda_t \approx 5.7$)
90	0.359	0.480	0.516	0.529	0.533	0.533	0.531	0.527	0.523	0.518	0.534 ($\lambda_t \approx 5.4$)
80	0.358	0.478	0.514	0.526	0.529	0.528	0.525	0.521	0.515	0.510	0.529 ($\lambda_t \approx 5.2$)
70	0.356	0.476	0.510	0.521	0.524	0.522	0.518	0.512	0.506	0.499	0.524 ($\lambda_t \approx 4.9$)
60	0.355	0.472	0.506	0.515	0.516	0.513	0.508	0.501	0.493	0.485	0.517 ($\lambda_t \approx 4.6$)
50	0.352	0.468	0.499	0.507	0.506	0.501	0.493	0.485	0.475	0.465	0.507 ($\lambda_t \approx 4.3$)
40	0.348	0.461	0.490	0.495	0.491	0.483	0.472	0.461	0.448	0.435	0.495 ($\lambda_t \approx 3.9$)
30	0.341	0.450	0.474	0.474	0.465	0.452	0.437	0.421	0.403	0.385	0.475 ($\lambda_t \approx 3.5$)
20	0.328	0.427	0.442	0.433	0.415	0.392	0.367	0.341	0.314	0.286	0.442 ($\lambda_t \approx 3.0$)
10	0.291	0.361	0.348	0.311	0.265	0.213	0.160	0.104	0.048		0.362 ($\lambda_t \approx 2.2$)
5	0.222	0.237	0.169	0.076							

The theoretical value of the power coefficient of ideal wind turbine can be calculated based on Formula (5.3). To facilitate lookup, use and analysis, the common scope and the data that is of theoretical significance has been listed in Table 5.1.

Now let's make preliminary prediction of the maximum power coefficient of actual wind turbine. The rightmost column in Table 5.1 contains the peak power and corresponding tip speed ratio which is obtained through analytic calculation of the lift drag ratio in the leftmost column. In Table 5.1, the maximum value of the power coefficient which has a lift drag ratio of less than 100 does not exceed 0.538 and is about 90% of Betz Limit. This result is achieved even with ideal blade structure and rather high supposed lift drag ratio of airfoil. It can be concluded that the power coefficient of the actual wind turbine which has a lift drag ratio of less than 100 in no way exceeds 0.538.

Being different from ideal wind turbine, actual wind turbine requires at least three considerations in addition to lift drag ratio. First, the power losses caused by limited number of blades shall be taken into consideration. The process of calculating the effect of limited number of blades on power coefficient is rather complex (see Section 9.1 for details), so only partial calculation results are provided here. For ideal blade structure, the changing trend of power coefficient after correction of tip loss with infinite lift drag ratio and a lift drag ratio of 100 is shown in Figure 5.5 which provides two curves of power coefficient in two different cases without considering the effect of limited number of blades to facilitate comparison. The curve ④ in Figure 5.5 suggests that only if the tip speed ratio falls within the range of 6-10 when the lift drag ratio is 100, will the power coefficient of the wind turbine with limited number of blades be a little bit higher than 0.500. If the lift drag ratio drops down to the practical zone within 100, there will be higher power loss.

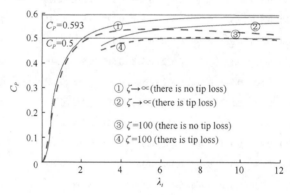

Figure 5.5　Schematic diagram of the effect of tip loss on power coefficient of ideal wind turbine

Second, ideal blade has quite complex structure and is difficult to process and manufacture, while the blade of actual wind turbine will certainly has simplified structure, which will further reduce the power coefficient.

Third, the airfoil of ideal blade remains unchanged, so the spanwise change of thickness is proportional to chord. For actual wind turbine, the thickness increases from tip to root and reaches up to about 100% of chord and even multiple airfoils are used, and in this case the irregular change of thickness has considerable impact on lift drag ratio, which will further reduce the power coefficient.

In addition, there will be some additional requirements for the shape of blade when considering the structural strength[38], vibration[39,40], pneumatic elastic deformation[41,42] and centrifugal rigidization[43,44] noise[45,46] and other aspects of the blade, and this contributes to further reduction of power coefficient.

The power loss caused by limited number of blades is inevitable; although it is not mandatory to simplify the structure of blade from theoretical view, the production process will certainly cost a lot, while thickness is changed to meet the requirements for structural strength, vibration and some other indicators. To summarize the above theoretical calculations and analysis of practical issues, it is fair to say that the power coefficient of the actual wind turbine which has a lift drag ratio of less 100 seldom exceeds 0.500 and it is about 84% of the Betz Limit.

Chapter 9 deals with detailed discussion about the calculation of tip loss caused by limited number of blades and derivation of the calculation formulas of highest performance and general performance of actual wind turbine.

5.2   General performance of torque

According to Formulas (5.2) and (4.6), the calculation formula of the torque coefficient of ideal wind turbine related to lift drag ratio and tip speed ratio is

$$C_M = \frac{16}{9}\int_0^1 \frac{x^2\left[\frac{2}{3}\zeta - \left(\lambda_t x + \frac{2}{9\lambda_t x}\right)\right]}{\left(\lambda_t x + \frac{2}{9\lambda_t x}\right)\zeta + \frac{2}{3}}\,dx$$

$$= \frac{16\left(-4 - 4\zeta^2 + 3\zeta\lambda_t + 3\zeta^3\lambda_t - 3\zeta^2\lambda_t^2\right)}{81\zeta^3\lambda_t^2} \tag{5.4}$$

$$- \frac{32\left(-2 - \zeta^2 + \zeta^4\right)}{243\zeta^4\lambda_t^3}\left[\ln(2\zeta) - \ln\left(2\zeta + 6\lambda + 9\zeta\lambda_t^2\right)\right]$$

$$+ \frac{64\left(-2 + \zeta^2 + 3\zeta^4\right)}{243\zeta^4\lambda_t^3\sqrt{-1 + 2\zeta^2}}\left(\mathrm{arccot}\sqrt{-1 + 2\zeta^2} - \arctan\frac{1 + 3\zeta\lambda_t}{\sqrt{-1 + 2\zeta^2}}\right)$$

This is the torque coefficient formula of ideal wind turbine operating in actual fluid.

As Formula (5.4) is rather complex, atlas and charts can be made with this formula to facilitate its use. See Figure 5.6 for the atlas that shows how C_M changes with the change of tip speed ratio λ_t.

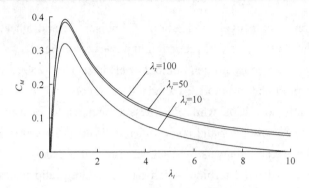

Figure 5.6 Atlas showing how the torque coefficient of ideal wind turbine changes with change of design tip speed ratio

The theoretical value of the torque coefficient of ideal wind turbine can be calculated based on Formula (5.4). To facilitate lookup, use and analysis, the common scope and the data that is of theoretical significance have been listed in Table 5.2.

Table 5.2 The theoretical value of torque coefficient formula of ideal wind turbine operating in actual fluid

Lift drag ratio ζ	Design tip speed ratio λ_t									
	1	2	3	4	5	6	7	8	9	10
∞	0.368	0.248	0.179	0.139	0.114	0.096	0.083	0.073	0.065	0.058
1000	0.367	0.247	0.179	0.139	0.113	0.095	0.082	0.072	0.064	0.058
900	0.367	0.247	0.179	0.139	0.113	0.095	0.082	0.072	0.064	0.058
800	0.367	0.247	0.179	0.139	0.113	0.095	0.082	0.072	0.064	0.058
700	0.367	0.247	0.178	0.138	0.113	0.095	0.082	0.072	0.064	0.058
600	0.367	0.247	0.178	0.138	0.113	0.095	0.082	0.072	0.064	0.057
500	0.366	0.246	0.178	0.138	0.112	0.094	0.081	0.071	0.064	0.057
400	0.366	0.246	0.178	0.138	0.112	0.094	0.081	0.071	0.063	0.057
300	0.365	0.245	0.177	0.137	0.111	0.094	0.081	0.071	0.063	0.056
200	0.364	0.244	0.176	0.136	0.110	0.093	0.080	0.070	0.062	0.055
100	0.360	0.241	0.173	0.133	0.107	0.090	0.077	0.067	0.059	0.052
90	0.359	0.240	0.172	0.132	0.107	0.089	0.076	0.066	0.058	0.052
80	0.358	0.239	0.171	0.131	0.106	0.088	0.075	0.065	0.057	0.051
70	0.356	0.238	0.170	0.130	0.105	0.087	0.074	0.064	0.056	0.050
60	0.355	0.236	0.169	0.129	0.103	0.086	0.073	0.063	0.055	0.048
50	0.352	0.234	0.166	0.127	0.101	0.083	0.070	0.061	0.053	0.046
40	0.348	0.230	0.163	0.124	0.098	0.080	0.067	0.058	0.050	0.044
30	0.341	0.225	0.158	0.118	0.093	0.075	0.062	0.053	0.045	0.039
20	0.328	0.213	0.147	0.108	0.083	0.065	0.052	0.043	0.035	0.029
10	0.291	0.181	0.116	0.078	0.053	0.036	0.023	0.013	0.005	

5.3 General performance of lift

According to Formulas (5.2) and (4.8), the calculation formula of the lift coefficient of ideal wind turbine related to lift drag ratio and tip speed ratio is

$$C_F = \frac{16}{9}\int_0^1 \frac{x\left[\frac{2}{3}\varsigma - \left(\lambda_t x + \frac{2}{9\lambda_t x}\right)\right]}{\left(\lambda_t x + \frac{2}{9\lambda_t x}\right)\varsigma + \frac{2}{3}}dx$$

$$= \frac{8(4+4\varsigma^2-3\varsigma\lambda_t)}{27\varsigma^2\lambda_t} + \frac{32(1+\varsigma^2)}{81\varsigma^3\lambda_t^2}\left[\ln(2\varsigma) - \ln\left(2\varsigma + 6\lambda_t + 9\varsigma\lambda_t^2\right)\right] \quad (5.5)$$

$$+ \frac{64(-1+\varsigma)(1+\varsigma)(1+\varsigma^2)}{81\varsigma^3\lambda_t^2\sqrt{-1+2\varsigma^2}}\left(\arctan\frac{1}{\sqrt{-1+2\varsigma^2}} - \arctan\frac{1+3\varsigma\lambda_t}{\sqrt{-1+2\varsigma^2}}\right)$$

This is the lift coefficient formula of ideal wind turbine operating in actual fluid.

As Formula (5.5) is rather complex, atlas and charts can be made with this formula to facilitate its use. See Figure 5.7 for the atlas that shows how C_F changes with the change of tip speed ratio λ_t.

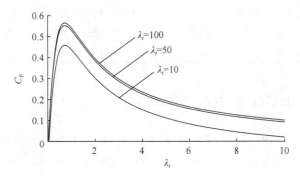

Figure 5.7 Atlas showing how the lift coefficient of ideal wind turbine changes with change of design tip speed ratio

The theoretical value of the lift coefficient of ideal wind turbine can be calculated based on Formula (5.5). To facilitate lookup, use and analysis, the common scope and the data that is of theoretical significance have been listed in Table 5.3.

Table 5.3 The theoretical value of lift coefficient formula of ideal wind turbine operating in actual fluid

Lift drag ratio ζ	Design tip speed ratio λ_t									
	1	2	3	4	5	6	7	8	9	10
∞	0.554	0.406	0.307	0.246	0.204	0.174	0.152	0.135	0.121	0.110
1000	0.552	0.404	0.306	0.245	0.203	0.173	0.151	0.134	0.120	0.109
900	0.552	0.404	0.306	0.244	0.203	0.173	0.151	0.134	0.120	0.109
800	0.552	0.404	0.306	0.244	0.203	0.173	0.151	0.134	0.120	0.109
700	0.552	0.404	0.306	0.244	0.203	0.173	0.151	0.134	0.120	0.109
600	0.552	0.404	0.306	0.244	0.202	0.173	0.151	0.133	0.120	0.108
500	0.551	0.403	0.305	0.244	0.202	0.173	0.150	0.133	0.119	0.108
400	0.551	0.403	0.305	0.243	0.202	0.172	0.150	0.133	0.119	0.108
300	0.550	0.402	0.304	0.242	0.201	0.171	0.149	0.132	0.118	0.107
200	0.548	0.400	0.302	0.241	0.199	0.170	0.148	0.130	0.117	0.105
100	0.541	0.395	0.297	0.236	0.195	0.165	0.143	0.126	0.112	0.101
90	0.540	0.394	0.296	0.235	0.194	0.164	0.142	0.125	0.111	0.100
80	0.538	0.392	0.295	0.233	0.192	0.163	0.141	0.124	0.110	0.099
70	0.536	0.390	0.293	0.232	0.191	0.161	0.139	0.122	0.108	0.097
60	0.533	0.388	0.290	0.229	0.188	0.159	0.137	0.120	0.106	0.095
50	0.529	0.384	0.287	0.226	0.185	0.156	0.134	0.117	0.103	0.092
40	0.523	0.379	0.282	0.221	0.180	0.151	0.129	0.112	0.098	0.087
30	0.513	0.370	0.274	0.213	0.173	0.143	0.122	0.105	0.091	0.080
20	0.494	0.352	0.258	0.198	0.157	0.128	0.106	0.089	0.076	0.065
10	0.437	0.302	0.210	0.151	0.111	0.083	0.061	0.045	0.031	0.020

5.4 General performance of thrust

Formula (4.10) suggests that the theoretical value of thrust coefficient formula of ideal wind turbine operating in actual fluid is 8/9 regardless of tip speed ratio and lift to drag ratio. See Chapter 9 for the thrust performance for which the effect of limited number of blades (leading to tip loss) is taken into consideration.

5.5 Summary of this chapter

Supposing that ideal wind turbine operates in actual fluid environment and its lift drag ratio is no longer infinite but a limited value, this operating environment is more approximate to actual environment. This chapter deals with the research on this state. The chord formula of ideal blade is substituted into the analytic expression of blade element performance and integrated to derive the calculation formulas of general performance of power, torque, lift and thrust coefficient. The study shows that the power, torque and lift performance of ideal wind turbine are irrelevant to the absolute values of airfoil and drag coefficient but is closely related to their ratio. The study also shows that the performance of ideal wind turbine is only related to two parameters—lift drag ratio and tip speed ratio, and is irrelevant to any other factor.

Chapter 6 Flat Airfoil Wind Turbine and Its Performance

Flat airfoil may be deemed as an airfoil of which both curvature and thickness are close to 0 and it is the simplest airfoil. The wind turbine which is composed of flat airfoil is the simplest wind turbine. The study on flat airfoil wind turbine and its performance may be deemed as a specific embodiment of the in-depth study on the general performance of ideal wind turbine.

This chapter contains systematic exploration of the lift and drag characteristics and their relations of flat airfoil at large and small angle of attack and deals with the derivation of the performance of flat airfoil wind turbine in ideal fluid environment. Besides, the blade airfoil of wind turbine may be simplified to flat airfoil to facilitate calculation under some special circumstances, and some analysis and calculation examples will be provided in this chapter.

6.1 Flat airfoil and performance estimation

In engineering practice, it is common that streamline airfoil is adopted instead of airfoil to minimize the drag. But the streamline shape of airfoil is rather complex and makes the calculation quite difficulty. Furthermore, as for actual fluid, the around flow lift and drag of streamline airfoil are different from that of ideal fluid, in particular the drag is closely related to viscosity and flow regime, so it is very difficult to do analytic calculation and it requires prolonged exploration to work out the solution. Although flat airfoil is seldom used in practice, the simplicity of flat structure makes the analytic calculation easier. If the around flow lift and drag formulas of flat airfoil are available, then the around flow regularity of flat airfoil can be referenced to explore approximately the general trend of the around flow of actual airfoil for easier analysis of problems. Replacing streamline airfoil with flat airfoil is in fact to focus on the principal contradiction while ignoring the minor contradiction temporarily to gain understanding of the complex problems.

6.1.1 Brief introduction to the around flow lift at small angle of attack

The around flow of flat airfoil at small angle of attack does not cause separation from the body, and it will generate lift no matter it is ideal fluid or viscous actual fluid.

Numerous literatures provide the lift derivation process and it is shown that the result is related to angle of attack and in the state of small angle of attack the lift coefficient is about $2\pi\sin\alpha$ which may be called classic lift coefficient formula. The typical derivation process of classic theory is as follows[47].

Coordinate system is shown in Figure 6.1, and here the chord of flat airfoil is set to C. The horizontal flat airfoil is exposed to the force of incoming flow which form an angle α with it.

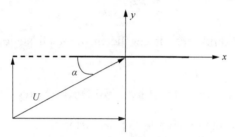

Figure 6.1 Calculation of around flow lift at small angle of attack

On the plane of $z(x, y)$, setting the velocity at an infinite point is U, the velocity near the flat airfoil is

$$u = U\cos\alpha + iU\sin\alpha \tag{6.1}$$

The complex potential of circulation flow around the flat airfoil is

$$w = zU\cos\alpha - iU\sin\alpha\sqrt{z^2 - \left(\frac{C}{2}\right)^2} + i\frac{\Gamma}{2\pi}\ln\left(z + \sqrt{z^2 - \left(\frac{C}{2}\right)^2}\right) \tag{6.2}$$

The complex velocity of this circulation flow is

$$\bar{v} = \frac{dw}{dz} = U\cos\alpha - i\frac{U\sin\alpha z}{\sqrt{z^2 - \left(\frac{C}{2}\right)^2}} + i\frac{\Gamma}{2\pi}\frac{1}{\sqrt{z^2 - \left(\frac{C}{2}\right)^2}}$$

$$= U\cos\alpha - i\frac{2\pi zU\sin\alpha - \Gamma}{2\pi\sqrt{z^2 - \left(\frac{C}{2}\right)^2}} \tag{6.3}$$

where C is the chord of flat airfoil, and Γ is circular rector. To ensure this formula is meaningful, the conditions near trailing edge point of flat airfoil shall fulfill Kutta-Joukowsky theorem, namely the velocity should be limited value, so the numerator must be set as 0, that is

$$2\pi z U \sin \alpha - \Gamma \big|_{z=\frac{C}{2}} = 0 \tag{6.4}$$

Therefore

$$\Gamma = \pi C U \sin \alpha \tag{6.5}$$

Lift coefficient

$$C_L = \frac{\rho U \Gamma}{\frac{1}{2}\rho U^2 C} = 2\pi \sin \alpha \tag{6.6}$$

For viscous actual fluid, the lift coefficient at small angle of attack is slightly smaller than this value.

6.1.2 Exploration of formula of around flow at large angle of attack

Ideal fluid does not impose drag to the flat airfoil in fluid field, and this is called D'Alembert paradox. Different from ideal fluid, viscous fluid imposes pressure on large angle of attack flat airfoil. The pressure equals to the differential pressure between that on the incident flow side and back pressure side of the flat airfoil; the direction of total pressure is perpendicular to flat airfoil and expressed with the symbol N. The total pressure may be decomposed into horizontal component (drag component) N_D and vertical component (lift component) N_L, as shown in Figure 6.2.

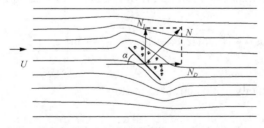

Figure 6.2 Flow graph of large angle of attack around flow of flat airfoil

Setting C as the chord of flat airfoil, obviously the total pressure $N(\alpha)$ imposed by the fluid on flat airfoil is proportional to the relative incident flow area ($C \sin \alpha / C$) of flat airfoil, and the larger the relative incident flow area is, the larger the pressure is; contrariwise it is smaller[48]. Suppose the functional relationship between total pressure and relative incident flow area is

$$N(\alpha) = a\left(\frac{C\sin\alpha}{C}\right)^b = a\sin^b\alpha \qquad (6.7)$$

where, a and b are constants.

When $\alpha = \pi/2$

$$N\left(\frac{\pi}{2}\right) = a\sin^b\frac{\pi}{2} = a \qquad (6.8)$$

It has been proven by a number of experiment data that, when the angle of attack is $\pi/2$ and Reynolds number falls within 10^4-10^6, the differential pressure drag coefficient $C_{\pi/2}$ of the two-dimensional flat airfoil which is perpendicular to the flow direction is 1.98[49,50], 2.01[51] or 2.06[52]. The difference of results may be related to experimental error. To simplify the expression of drag coefficient, set $C_{\pi/2}$ as 2, and the differential pressure drag on vertical flat airfoil is

$$N\left(\frac{\pi}{2}\right) = C_{\pi/2} \cdot \frac{1}{2}\rho U^2 C = \rho U^2 C \qquad (6.9)$$

Compared with Formula (6.8), we get

$$a = \rho U^2 C \qquad (6.10)$$

Obviously a is a constant for given incoming flow field and chord of flat airfoil. Substitute it into Formula (6.7) to get the total pressure

$$N(\alpha) = \rho U^2 C \sin^b\alpha \qquad (6.11)$$

The total pressure coefficient is

$$C_N = \frac{N(\alpha)}{\frac{1}{2}\rho U^2 C} = 2\sin^b\alpha \qquad (6.12)$$

Now the left problem is to determine the coefficient b. The first step is to demonstrate that coefficient b can only be odd. The lift component of total pressure is

$$N_L(\alpha) = N(\alpha)\cos\alpha = \rho U^2 C \sin^b\alpha \cos\alpha \qquad (6.13)$$

The lift component coefficient is

$$C_{NL}(\alpha) = 2\sin^b\alpha \cos\alpha \qquad (6.14)$$

The sizes of lift component coefficient when the angle of attack is α and $-\alpha$ respectively are certainly equal but in opposite directions due to the symmetry of the flat airfoil, and this can be expressed with the formula below

$$C_{NL}(\alpha) = -C_{NL}(-\alpha) \tag{6.15}$$

or

$$\sin^b \alpha \cos \alpha = -\sin^b(-\alpha)\cos(-\alpha) \tag{6.16}$$

so

$$\sin^b \alpha \cos \alpha = -(-1)^b \sin^b \alpha \cos \alpha \tag{6.17}$$

To make this formula satisfied, it is obvious that the coefficient b can only be odd (it is assumed above that b is not negative), namely the value of b can only be 1, 3, 5, ⋯

Now let's discuss the trend of the influence of the change of value b on lift and drag coefficients. The drag component of total pressure is

$$N_D = N \sin \alpha = \rho U^2 C \sin^{b+1} \alpha \tag{6.18}$$

The drag component coefficient is

$$C_{ND}(\alpha) = 2\sin^{b+1} \alpha \tag{6.19}$$

When b equals to 1, 3 and 5 respectively, the corresponding lift and drag coefficient curves are shown in Figure 6.3.

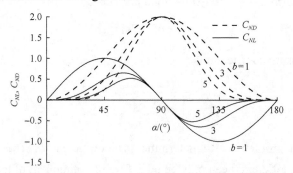

Figure 6.3   The curve showing how large angle of attack around flow lift and drag component coefficients change with the change of coefficient b

According to Figure 6.3, no matter what the value of b is, the maximum value of drag component coefficient is 2 when the flat airfoil is perpendicular to the incoming flow. The absolute value of lift component coefficient increases as the value of b decreases at any angle of attack expect for 0. Derive the lift component coefficient and set it as 0.

$$\frac{\partial C_{NL}(\alpha)}{\partial \alpha} = b\cos^2\alpha \sin^{b-1}\alpha - \sin^{b+1}\alpha = 0 \tag{6.20}$$

The angle of attack when extreme value is applied within [0°, 90°] can be obtained

$$\alpha_f = \operatorname{arc cot}\frac{1}{\sqrt{b}} \tag{6.21}$$

At this moment the lift component coefficient is

$$C_{NL\max}(\alpha) = 2\sin^b\alpha_f \cos\alpha_f = \frac{2\sqrt{b^b}}{\sqrt{(1+b)^{b+1}}} \tag{6.22}$$

Table 6.1 provides the maximum value of lift component coefficient and corresponding angle of attack when the coefficient b is set as an odd within the range from 1 to 9.

Table 6.1　The change of maximum lift coefficient and corresponding angle of attack along the coefficient b

Coefficient b	The maximum value of lift component coefficient $C_{NL\max}$	The angle of attack when lift component coefficient reaches up to maximum value $\alpha_f/(°)$
1	1	45
3	0.650	60
5	0.518	65.9
7	0.443	69.3
9	0.394	71.6

It is shown in Table 6.1 that the maximum value of lift component coefficient appears when $b=1$ angle of attack is 45° and at this moment both lift coefficient and drag coefficient are 1. Obviously $b=1$ is a very special constant and its consistency with experimental result shall be emphasized. First of all, let's take a look at the situations when $b=1$. It is derived from Formulas (6.12), (6.14) and (6.19) that the around flow total pressure and its lift and drag component coefficient expressions respectively are

$$C_N = 2\sin\alpha \tag{6.23}$$

$$C_{NL} = \sin 2\alpha \tag{6.24}$$

$$C_{ND} = 2\sin^2 \alpha \qquad (6.25)$$

It is possible to draw the theoretical curves of lift and drag coefficients of total pressure when the angle of attack of flat airfoil is 0°-180° (Figure 6.4).

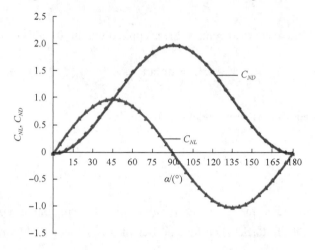

Figure 6.4 Curve of around flow lift and drag component coefficient of flat airfoil at 0°-180°

AVIC Aerodynamics Research Institute has done experiment of two-dimensional airfoil of wind turbine at large angle of attack[53]. What is used in experiment is NACA 0015 symmetrical airfoil and the Reynolds number is 0.5×10^6. The curve of experiment test data is shown in Figure 6.5 which also provides the theoretical curves calculated based on Formulas (6.24) and (6.25) in this book.

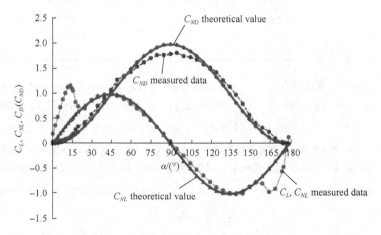

Figure 6.5 Curve of the measured value and calculated value of lift and drag coefficient of symmetrical airfoil NACA 0015

Figure 6.6 shows the curve that compares the measurement of symmetrical airfoil NACA 0012 with a Reynolds number of 0.5×10^6 provided by Sandia National Lab[54] with the theoretical value based on Formulas (6.24) and (6.25) in this book.

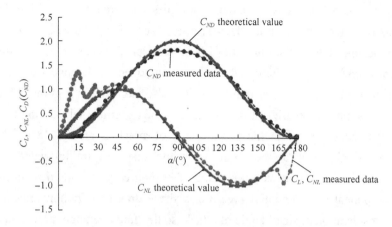

Figure 6.6 Curve of the measured value and calculated value of lift and drag coefficient of symmetrical airfoil NACA 0012

The experimental data in Figures 6.5 and 6.6 shows that the maximum measured value of the lift component coefficient of symmetrical airfoil within the range of large angle of attack is about 1 and close to the value calculated when $b=1$. As shown in Table 6.1, when the value of b is greater than or equal to 3, the maximum value of lift component coefficient is less than or equals to 0.65 which differs too much from and denied by the experimental result, indicating that the value of coefficient b can only be 1. Thus it can be determined that Formulas (6.23)-(6.25) are the expressions of large angle of attack around flow total pressure of flat airfoil, its lift and drag respectively.

The calculated value differs from measured value under large angle of attack mainly due to: ①the difference between the airfoil used for experiment and flat airfoil in terms of asymmetry between front and back and non-uniform thickness contributes a large portion to the error. ②The calculated value of drag coefficient curve is higher than measured value when the angle of attack is around 90° because the leading edge of airfoil is different from flat shape and the streamline structure delays the separation of flow while reduces the drag. The formulas in this book are based on the drag coefficient formula of vertical plate airfoil, so there will be no significant error if flat airfoil is used for the experiment.

There is no comparability between the curve of lift component coefficient formula and the curve of small angle of attack lift coefficient because circular rector is dominate in the state of small angle of attack. In the state of large angle of attack, the condition in which the circular rector exists disappear and the differential pressure caused by viscosity becomes dominant, and lift and drag are nothing but the components of differential pressure. However, it is shown in Figures 6.5 and 6.6 that the lift component may be the minimum lift in stall zone, because total pressure also exists in stall zone and the differential pressure component will support the lift when the circular rector lift is no longer dominant. It is also indicated in Figures 6.5 and 6.6 that the drag component coefficient is also close to the experimental data in the case of small angle of attack, in particular if the angle of attack is around 180° when the trailing edge of airfoil is at the front while the shape of trailing edge is more approximate to flat compared with the shape of leading edge. It should be noted that, the drag component of differential pressure is shape drag and friction drag must be taken into consideration at small angle of attack, so the drag component formula cannot be directly used for calculation of the drag at small angle of attack.

Formulas (6.23)-(6.25) are approximate formulas obtained through analysis and deduction based on experimental results. These formulas may be used for approximate calculation of the around flow of flat airfoil at large angle of attack, its lift component and drag component, and may also be used for preliminary estimate of the around flow of systematical airfoil with an applicable Reynolds number of 10^4-10^6. Although what are obtained herein are approximate formulas, the numerous experimental data of the around flow at large angle of attack may be expressed with a group of simple analytic formulas to facilitate the analysis and preliminary estimate of the power, thrust, torque and other performance factors of wind turbine and other fluid-driven machinery, which can significantly save time. It is shown in Figures 6.5 and 6.6 that the drag coefficient given by the formula in this book is more coincident with experimental data; the total pressure lift component coefficient at large angle of attack is more approximate to measured value.

As the differential pressure drag formula of incompressible steady flow two-dimensional flat airfoil is used in formula deduction process, so the scope of application of total pressure and its component formula is: ①fluid is incompressible; ②two-dimensional steady flow; ③viscosity is dominant (including the generation of off-flow); ④Reynolds number of 10^4-10^6.

6.1.3 Drag coefficient of around flow at small angle of attack

The drag coefficient of flat airfoil at small angle of attack is hereby researched based on the actual conditions of flow. Formula (6.25) reflects how the horizontal component of total pressure changes with the change of angle of attack, and as this component is related to the angle of attack of flat airfoil and thus related to the shape of airfoil in flow field, it can be called shape drag. Formula (6.25) does not reflect the friction drag that is perpendicular to the flat surface within the boundary layer. The shape drag is 0 when the angle of attack is 0, but the total drag (shape drag + friction drag) cannot be 0 and the friction drag shall be taken into consideration. In fact, the friction drag always exists at small angle of attack before separation of the flow and its value changes within the range of small angle of attack and may be deemed as a constant. In this book, $2C_f$ represents friction drag coefficient at zero angle of attack and C_D represents total drag, thus the total drag coefficient within small angle of attack is

$$C_D = 2C_f + 2\sin^2 \alpha \tag{6.26}$$

Friction drag coefficient $2C_f$ can be calculated using boundary layer theory based on the state of laminar flow or turbulent flow. As the absolute value of $2C_f$ is very small and usually falls within the range of 0.003-0.05, it is difficult to find it out directly on the graph. Although the size of measurement is very small, it should not be ignored because it is equivalent to the shape drag at small angle of attack and is a significant factor that affects the lift drag ratio of airfoil. When the angle of attack increases after the separation of flow, the friction drag is relatively reduced while the shape drag increases rapidly, and as the ratio of these two factors is very small, the friction drag is negligible; therefore, there is no need to correct the Formula (6.25) at large angle of attack.

6.1.4 The functional relationship between lift and drag

Both lift and drag are functions of the angle of attack, so the angle of attack may be deemed as a parameter and the functional relationship between lift and drag of flat airfoil could be obtained by eliminating the angle of attack[55].

1) Relationship between around flow lift and drag at small angle of attack

Formulas (6.6) and (6.25) for lift and drag at small angle of attack derive the formula below after eliminating the angle of attack:

$$C_D = 2C_f + 2\sin^2\alpha = 2C_f + 2\left(\frac{C_L}{2\pi}\right)^2 \tag{6.27}$$

Formula (6.27) may also be expressed as follows:

$$C_L^2 = 2\pi^2\left(C_D - 2C_f\right) \tag{6.28}$$

Formula (6.28) indicates that the lift and drag curve is a parabola on which the peak is at $(2C_f, 0)$ and focus is at $(\pi^2/2, 0)$ at small angle of attack. See Figure 6.7 for the shape of curve.

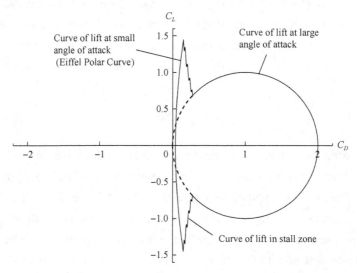

Figure 6.7   Diagram of relationship between around flow lift and drag at full angle of attack

The lift and drag curve is often referred to as the Eiffel Polar Curve. The tangent point where the straight line crossing the origin point is tangent to the Eiffel Polar Curve witnesses the highest lift-drag ratio, and it can be proven that the corresponding angle of attack is the optimal angle of attack (Section 3.2).

2) Relationship between around flow lift and drag at large angle of attack

Formulas (6.24) and (6.25) for lift and drag at large angle of attack derive the formula below after eliminating the angle of attack:

$$\begin{aligned}C_{NL}^2 &= \sin^2 2\alpha = 4\sin^2\alpha\left(1-\sin^2\alpha\right) \\ &= 2C_{ND}\left(1-C_{ND}/2\right) = C_{ND}\left(2-C_{ND}\right)\end{aligned} \tag{6.29}$$

Formula (6.29) may also be expressed as follows:

$$C_{NL}^2 + (C_{ND} - 1)^2 = 1 \tag{6.30}$$

Formula (6.30) shows that the lift-drag curve is a circle which has its center at (1,0) and has radius of 1 at large angle of attack.

The shape of the lift coefficient curve on which drag coefficient is the abscissa at large and small angle of attack is shown in Figure 6.7.

6.1.5 The regularity of change of lift and drag

The regularity of change of around flow lift and drag component of airfoil is summarized as follows based on the results of analysis and derivation.

(1) Without considering the effect of circular rector, the horizontal component of total pressure is the dominant drag beyond small angle of attack and its vertical component is the dominant lift beyond small angle of attack due to the effect of viscosity.

(2) The circular rector lift fluctuates violently in the transition zone between large and small angle of attack, but the total lift is supported by the lift component of pressure and thus will not drop down to 0. Lift component is the minimum value of the total lift of any angle of attack.

(3) The around flow pressure, lift and drag coefficients of flat airfoil at full angle of attack can be calculated with the dimensionless formula in Table 6.2.

Table 6.2 Calculation formulas of around flow lift and drag coefficients at full angle of attack

Formulas	Scope of application	Flow characteristics
$C_L = 2\pi \sin \alpha$	Estimation of lift at small angle of attack	Flow is not off from body, and lift is dominated by circular rector
$C_{NL} = \sin 2\alpha$	Estimation of lift at large angle of attack	Flow is off from body, and lift is vertical component of total pressure
$C_{L\max} = 2\pi \sin \alpha$ $C_{L\min} = \sin 2\alpha$	Estimation of lift in stall zone	Stall zone in which flow starts to separate from body until completely separated, and lift fluctuates violently
$C_D = 2C_f + 2\sin^2 \alpha$	Estimate of drag at small angle of attack	Flow is not separated from body, and magnitude of friction drag is equivalent to that of shape drag
$C_{ND} = 2\sin^2 \alpha$	Estimate of drag at large angle of attack	Flow is separated from body, and drag is horizontal component of total pressure
$C_{D\max} = 2C_f + 2\sin^2 \alpha$ $C_{D\min} = 2\sin^2 \alpha$	Estimation of drag in stall zone	Flow starts to separate from body until completely separated, drag increases gradually with little fluctuation

The friction drag coefficient $2C_f$ at zero angle of attack is a constant and easy to measure, so that there are determined expressions of all lift and drag coefficients, which is a very favorable condition for differential and integral operation. This set of formulas provide powerful analysis means for studying the trend of the torque, power, lift and other of macro performance indicators of wind turbine and other blade-based fluid machinery.

6.1.6 Lift drag ratio of flat airfoil

The lift drag ratio of flat airfoil at small angle of attack is derived from Formulas (6.6) and (6.27).

$$\zeta = \frac{C_L(\alpha)}{C_D(\alpha)} = \frac{2\pi \sin\alpha}{2C_f + 2\sin^2\alpha} \tag{6.31}$$

Hereby it is possible to draw the curve of relationship between lift drag ratio ζ and angle of attack α of flat airfoil, as shown in Figure 6.8.

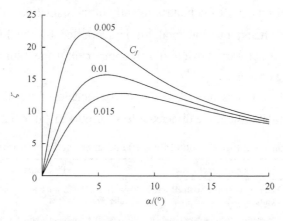

Figure 6.8 Relationship between lift drag ratio and angle of attack of flat airfoil

The lift drag ratio of the optimal angle of attack is about 15.71 when $C_f = 0.01$.

6.2 Structure of ideal blade with flat airfoil

6.2.1 The optimal angle of attack

The optimal angle of attack is the one that results in maximum blade element lift drag ratio. The blade runs at small angle of attack in design situation, and substitute Formulas (6.6) and (6.27) of lift coefficient and drag coefficient of flat airfoil at small angle of attack into Formula (3.10).

$$\frac{2\pi \sin \alpha}{2C_f + 2\sin^2 \alpha} = \frac{2\pi \cos \alpha}{4\sin \alpha \cos \alpha} \tag{6.32}$$

Obtain

$$\sin \alpha_b = \sqrt{C_f} \text{ or } \alpha_b = \arcsin \sqrt{C_f} \tag{6.33}$$

This is the formula of optimal angle of attack of flat airfoil. We reach an important conclusion thereby: the sinusoidal value of the optimal angle of attack of flat airfoil is the square foot of half of friction drag coefficient when the blade has zero angle of attack. This reveals the regularity of how the optimal angle of attack of flat airfoil changes along with the friction drag coefficient, as shown in Figure 6.9.

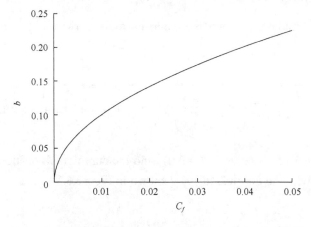

Figure 6.9 The relationship between the optimal angle of attack and friction drag coefficient of flat airfoil

It is also possible to obtain the formulas of lift and drag coefficients when blade element has the optimal angle of attack.

$$C_L(\alpha_b) = 2\pi \sin \alpha_b = 2\pi \sqrt{C_f} \tag{6.34}$$

$$C_D(\alpha_b) = 2C_f + 2\sin^2 \alpha_b = 4C_f \tag{6.35}$$

6.2.2 Ideal twist

Twist and angle of attack is mutually complementary and their sum equals to the inflow angle, so the twist that fulfills the optimal angle of attack is ideal twist. For flat airfoil, the specific expression of twist is as follows according to Formulas (3.13) and (6.33).

$$\beta = \arctan\frac{6\lambda_t(r/R)}{9\lambda_t^2(r/R)^2+2} - \arcsin\sqrt{C_f} \quad (6.36)$$

It is shown in the above formula that the ideal twist of flat airfoil is determined by tip speed ratio and friction drag coefficient, and twist is closely related to friction drag coefficient. This indicates that the fineness of blade will decrease gradually while the friction drag coefficient will increase gradually over time, so the ideal twist will change, and this shall be taken into consideration in design phase.

6.2.3 Ideal chord

Substitute Formulas (6.34) and (6.35) into Formula (3.18) to obtain the ideal chord of flat blade.

$$\frac{C}{R} = \frac{8\pi}{9B}\frac{x}{\left[\pi\sqrt{C_f}\left(\lambda_t x + \frac{2}{9\lambda_t x}\right) + \frac{4}{3}C_f\right]\sqrt{\left(\lambda_t x + \frac{2}{9\lambda_t x}\right)^2 + \left(\frac{2}{3}\right)^2}} \quad (6.37)$$

Set the number of blades $B=3$, $C_f=0.01$, and the curve graph of ideal chord of flat airfoil expressed with the Formula (6.37) is shown in Figure 6.10.

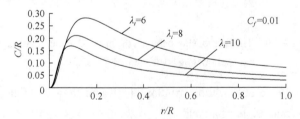

Figure 6.10 Curve graph of ideal chord of flat airfoil

6.3 Performance of flat airfoil wind turbine

Formulas (2.8) - (2.17) derived above are applicable to all airfoils. But there are different forms of expression for lift and drag coefficients due to the different of airfoil.

More specific formula of the pneumatic performance of the simplest flat airfoil blade (namely zero curvature and zero thickness airfoil blade) in design situation is provided here.

The blade runs at small angle of attack in design situation, and the theoretical lift coefficient of flat airfoil at small angle of attack has been given in Formula (6.6).

$$C_L = 2\pi \sin \alpha \tag{6.38}$$

Formula (2.26) provides the lift coefficient of flat airfoil at small angle of attack:

$$C_D = 2C_f + 2\sin^2 \alpha \tag{6.39}$$

where $2C_f$ is the friction drag coefficient of the blade at zero angle of attack; the friction drag coefficient at small angle of attack changes invisibly and may be deemed as constant. Where: the second item on right side of the equal sign is the shape drag coefficient of flat airfoil.

Substitute Formulas (6.38) and (6.39) of the lift and drag coefficients of flat airfoil into Formulas (2.10)-(2.13) respectively to obtain the blade element performance formula of flat airfoil.

$$dT = \frac{1}{2}\rho U^2 C \left[2\pi \left(\lambda + \frac{2}{9\lambda} \right) \sin \alpha + \frac{4}{3} \left(C_f + \sin^2 \alpha \right) \right] \sqrt{\left(\lambda + \frac{2}{9\lambda} \right)^2 + \left(\frac{2}{3} \right)^2} \, dr \tag{6.40}$$

$$dF = \frac{1}{2}\rho U^2 C \left[\frac{4}{3}\pi \sin \alpha - 2\left(C_f + \sin^2 \alpha \right)\left(\lambda + \frac{2}{9\lambda} \right) \right] \sqrt{\left(\lambda + \frac{2}{9\lambda} \right)^2 + \left(\frac{2}{3} \right)^2} \, dr \tag{6.41}$$

$$dM = \frac{1}{2}\rho U^2 C \left[\frac{4}{3}\pi \sin \alpha - 2\left(C_f + \sin^2 \alpha \right)\left(\lambda + \frac{2}{9\lambda} \right) \right] \sqrt{\left(\lambda + \frac{2}{9\lambda} \right)^2 + \left(\frac{2}{3} \right)^2} \, r \, dr \tag{6.42}$$

$$dP = \frac{1}{2}\rho U^3 C \lambda \left[\frac{4}{3}\pi \sin \alpha - 2\left(C_f + \sin^2 \alpha \right)\left(\lambda + \frac{2}{9\lambda} \right) \right] \sqrt{\left(\lambda + \frac{2}{9\lambda} \right)^2 + \left(\frac{2}{3} \right)^2} \, dr \tag{6.43}$$

Similar to the above derivation process, the performance of the wind turbine which is composed of B flat airfoil blades can be obtained by integrating the blade element performance formula.

$$C_T = \frac{B}{\pi}\int_0^1 \left(\frac{C}{R}\right)\left[2\pi\left(\lambda_\gamma x + \frac{2}{9\lambda_\gamma x}\right)\sin\alpha + \frac{4}{3}\left(C_f + \sin^2\alpha\right)\right]\sqrt{\left(\lambda + \frac{2}{9\lambda_\gamma x}\right)^2 + \left(\frac{2}{3}\right)^2}\,dx$$

(6.44)

$$C_F = \frac{B}{\pi}\int_0^1 \left(\frac{C}{R}\right)\left[\frac{4}{3}\pi\sin\alpha - 2\left(C_f + \sin^2\alpha\right)\left(\lambda_\gamma x + \frac{2}{9\lambda_\gamma x}\right)\right]\sqrt{\left(\lambda_\gamma x + \frac{2}{9\lambda_\gamma x}\right)^2 + \left(\frac{2}{3}\right)^2}\,dx$$

(6.45)

$$C_M = \frac{B}{\pi}\int_0^1 x\left(\frac{C}{R}\right)\left[\frac{4}{3}\pi\sin\alpha - 2\left(C_f + \sin^2\alpha\right)\left(\lambda_\gamma x + \frac{2}{9\lambda_\gamma x}\right)\right]\sqrt{\left(\lambda_\gamma x + \frac{2}{9\lambda_\gamma x}\right)^2 + \left(\frac{2}{3}\right)^2}\,dx$$

(6.46)

$$C_P = \frac{B}{\pi}\int_0^1 \lambda_\gamma x\left(\frac{C}{R}\right)\left[\frac{4}{3}\pi\sin\alpha - 2\left(C_f + \sin^2\alpha\right)\left(\lambda_\gamma x + \frac{2}{9\lambda_\gamma x}\right)\right]\sqrt{\left(\lambda_\gamma x + \frac{2}{9\lambda_\gamma x}\right)^2 + \left(\frac{2}{3}\right)^2}\,dx$$

(6.47)

Theorem of momentum is used in the process of deriving the speed induction factor Formula (2.2), so all of the above formulas are constrained by the theorem of momentum, namely chord must match twist; otherwise, the calculation result may not be the result in optimal operation state (see Section 10.4 for the example of adjusting twist based on chord).

6.3.1 Power performance

According to Formulas (6.34) and (6.35), the lift coefficient and drag coefficients of flat airfoil at the optimal angle of attack are

$$C_L(\alpha_b) = 2\pi\sin\alpha_b = 2\pi\sqrt{C_f} \tag{6.48}$$

$$C_D(\alpha_b) = 2C_f + 2\sin^2\alpha_b = 4C_f \tag{6.49}$$

Substitute Formula (4.1) and do integration to obtain the formula that deals with how the power coefficient of flat airfoil changes with design tip speed ratio and friction coefficient in the optimal conditions:

$$C_P = \frac{16}{9} \int_0^1 \frac{\lambda_t x^2 \left[\frac{2}{3} C_L(\alpha_b) - \left(\lambda_t x + \frac{2}{9\lambda_t x}\right) C_D(\alpha_b)\right]}{\left(\lambda_t x + \frac{2}{9\lambda_t x}\right) C_L(\alpha_b) + \frac{2}{3} C_D(\alpha_b)} dx$$

$$= \frac{16}{9} \int_0^1 \frac{\lambda_t x^2 \left[\frac{2}{3} \cdot 2\pi\sqrt{C_f} - \left(\lambda_t x + \frac{2}{9\lambda_t x}\right) \cdot 4C_f\right]}{\left(\lambda_t x + \frac{2}{9\lambda_t x}\right) \cdot 2\pi\sqrt{C_f} + \frac{2}{3} \cdot 4C_f} dx$$

$$= \frac{64\sqrt{2C_f}\left(32C_f^2 - 4\pi^2 C_f - 3\pi^4\right)}{243\pi^4 \lambda_t^2 \sqrt{\pi^2 - 2C_f}} \left[\arctan\frac{\sqrt{2C_f}}{\sqrt{\pi^2 - 2C_f}} - \arctan\frac{2\sqrt{C_f} + 3\pi\lambda_t}{\sqrt{2\pi^2 - 4C_f}}\right]$$

$$+ \frac{16}{243\pi^4 \lambda_t^2}\left[\pi^4\left(9\lambda_t^2 + 2\ln 2\pi\right) - 96\pi\lambda_t C_f^{3/2} - 6\pi^3 \lambda_t \sqrt{C_f}\left(4 + 3\lambda_t^2\right) - 64 C_f^2 \ln 2\pi\right]$$

$$+ \frac{16}{243\pi^4 \lambda_t^2}\left[4\pi^2 C_f\left(9\lambda^2 - 2\ln 2\pi\right) + \left(64 C_f^2 + 8\pi^2 C_f - 2\pi^4\right)\ln\left(2\pi + 12\lambda_t \sqrt{C_f} + 9\pi\lambda_t^2\right)\right]$$

(6.50)

The atlas showing how C_P changes along λ_t and C_P along C_f can be drawn based on Formula (6.50). See Figures 6.11 and 6.12 respectively.

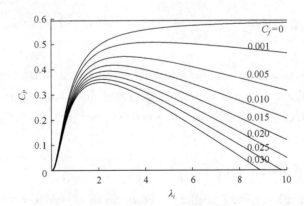

Figure 6.11 Atlas of the power coefficients of flat airfoil ideal wind turbine that correspond to different friction drag coefficients

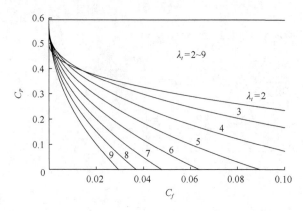

Figure 6.12 Atlas of the power coefficients of flat airfoil ideal wind turbine that correspond to different tip speed ratio

The horizontal line in Figures 6.11 and 6.12 is Betz Limit (0.593). This is the curve atlas of ideal wind turbine calculated on the basis of stable operation state (speed induction factor is a stable value), and the value of maximum power coefficient of any flat airfoil ideal wind turbine can be found once C_f and λ_t is given.

According to Formula (6.50), the power coefficient which changes along tip speed ratio reaches the maximum value when C_f approaches an infinite value.

$$C_{P\max} = \lim_{C_f \to 0} C_P = \frac{16}{243\lambda_t^2}\left[9\lambda_t^2 - 2\ln\left(9\lambda_t^2 + 2\right) + 2\ln 2\right] \qquad (6.51)$$

This formula is identical to the related limit Formula (4.3) of common power coefficient derived above and it is proven again that the related limit formula of tip speed ratio is irrelevant to airfoil when drag is 0, and this is the limit formula of wind turbine with any airfoil.

6.3.2 Torque performance

Substitute the lift and drag coefficient Formulas (6.48) and (6.49) that correspond to the optimal angle of attack of flat airfoil into Formula (4.6) to obtain the torque coefficient.

$$C_M = \frac{16}{9}\int_0^1 \frac{x^2\left[\frac{2}{3}C_L(\alpha_b) - \left(\lambda_t x + \frac{2}{9\lambda_t x}\right)C_D(\alpha_b)\right]}{\left(\lambda_t x + \frac{2}{9\lambda_t x}\right)C_L(\alpha_b) + \frac{2}{3}C_D(\alpha_b)} dx$$

$$= \frac{16}{9}\int_0^1 \frac{x^2\left[\frac{2}{3}\cdot 2\pi\sqrt{C_f} - \left(\lambda_t x + \frac{2}{9\lambda_t x}\right)\cdot 4C_f\right]}{\left(\lambda_t x + \frac{2}{9\lambda_t x}\right)\cdot 2\pi\sqrt{C_f} + \frac{2}{3}\cdot 4C_f} dx$$

$$= \frac{64\sqrt{2C_f}\left(32C_f^2 - 4\pi^2 C_f - 3\pi^4\right)}{243\pi^4 \lambda_t^3 \sqrt{\pi^2 - 2C_f}}\left(\arctan\frac{\sqrt{2C_f}}{\sqrt{\pi^2 - 2C_f}} - \arctan\frac{2\sqrt{C_f} + 3\pi\lambda_t}{\sqrt{2\pi^2 - 4C_f}}\right)$$

$$+ \frac{16}{243\pi^3 \lambda_t^3}\left[-96C_f^{3/2}\lambda_t + 36\pi C_f \lambda_t^2 - 6\pi^2 \sqrt{C_f}\lambda_t\left(4 + 3\lambda_t^2\right) + \pi^3\left(9\lambda_t^2 + 2\ln 2\pi\right)\right]$$

$$+ \frac{32\left(8C_f + \pi^2\right)}{243\pi^4 \lambda_t^3}\left[4C_f \ln 2\pi - \left(4C_f + \pi^2\right)\ln\left(9\pi\lambda_t^2 + 12\sqrt{C_f} + 2\pi\right)\right]$$

(6.52)

Figure 6.13 shows the trend of how the torque coefficient changes along tip speed ratio.

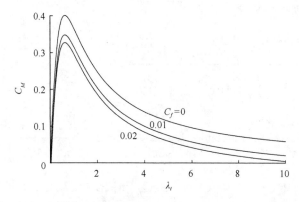

Figure 6.13 Curve showing how the torque coefficient of flat airfoil ideal wind turbine changes along tip speed ratio

It is indicated in the figures that drag and design tip speed ratio has considerable impact on torque, and for high speed wind turbine, the torque reduces rapidly as the drag and tip speed ratio increase.

According to Formula (6.52), the torque coefficient which changes along tip speed ratio reaches the maximum value when C_f approaches an infinite value.

$$C_{M\max} = \lim_{C_f \to 0} C_M = \frac{16}{243\lambda_t^3}\left[9\lambda_t^2 - 2\ln\left(9\lambda_t^2 + 2\right) + 2\ln 2\right] \tag{6.53}$$

This formula is identical to the related limit Formula (4.7) of common torque coefficient derived above and it is proven that the related limit formula of tip speed ratio is irrelevant to airfoil when drag is 0, and this is the limit formula of wind turbine with any airfoil.

6.3.3 Lift performance

Now let's explore the trend of how the lift coefficients of flat airfoil wind turbine change. Substitute the lift and drag coefficient Formulas (6.48) and (6.49) that correspond to the optimal angle of attack of flat airfoil into Formula (4.8) to obtain the lift coefficient of ideal wind turbine.

$$C_F = \frac{16}{9}\int_0^1 \frac{x\left[\frac{2}{3}C_L(\alpha_b) - \left(\lambda_t x + \frac{2}{9\lambda_t x}\right)C_D(\alpha_b)\right]}{\left(\lambda_t x + \frac{2}{9\lambda_t x}\right)C_L(\alpha_b) + \frac{2}{3}C_D(\alpha_b)}dx$$

$$= \frac{16}{9}\int_0^1 \frac{x\left[\frac{2}{3}\cdot 2\pi\sqrt{C_f} - \left(\lambda_t x + \frac{2}{9\lambda_t x}\right)\cdot 4C_f\right]}{\left(\lambda_t x + \frac{2}{9\lambda_t x}\right)\cdot 2\pi\sqrt{C_f} + \frac{2}{3}\cdot 4C_f}dx$$

$$= \frac{32\sqrt{2}\left(\pi^4 - 16C_f^2\right)}{81\pi^3\lambda_t^2\sqrt{\pi^2 - 2C_f}}\left(\arctan\frac{\sqrt{2C_f}}{\sqrt{\pi^2 - 2C_f}} - \arctan\frac{2\sqrt{C_f} + 3\pi\lambda_t}{\sqrt{2\pi^2 - 4C_f}}\right) \tag{6.54}$$

$$+\frac{16}{81\pi^3\lambda_t^2}\left[24\pi\lambda_t C_f - 4\sqrt{C_f}\left(\pi^2 + 4C_f\right)\ln\left(2\pi + 12\lambda_t\sqrt{C_f} + 9\pi\lambda_t^2\right)\right]$$

$$+\frac{16}{81\pi^3\lambda_t^2}\left[6\pi^3\lambda_t + 16C_f^{3/2}\ln 2\pi + \pi^2\sqrt{C_f}\left(4\ln 2 + 4\ln \pi - 9\lambda_t^2\right)\right]$$

Figure 6.14 shows the trend of how the lift coefficient changes along tip speed ratio.

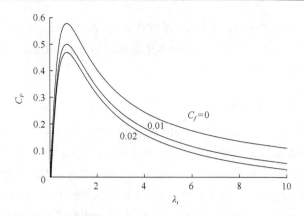

Figure 6.14　Curve showing how the lift coefficient of flat airfoil ideal wind turbine changes along tip speed ratio

Set friction drag coefficient $C_f = 0$, then

$$C_{F\max} = \lim_{C_f \to 0} C_F = \frac{32}{81\lambda_t^2}\left(3\lambda_t - \sqrt{2}\arctan\frac{3\lambda_t}{\sqrt{2}}\right) \quad (6.55)$$

This formula is identical to the related limit Formula (4.9) of common lift coefficient derived above and it is proven that the related limit formula of tip speed ratio is irrelevant to airfoil when drag is 0, and this is the limit formula of wind turbine with any airfoil.

6.3.4　Thrust performance

Now let's explore the trend of how the thrust coefficient of flat airfoil wind turbine changes. Substitute the lift and drag coefficients Formulas (6.48) and (6.49) that correspond to the optimal angle of attack of flat airfoil into Formula (2.14).

$$C_T = \frac{B}{\pi}\int_0^1 \left\{\frac{16\pi}{9B}\frac{x}{\left[\left(\lambda_t x + \frac{2}{9\lambda_t x}\right)\cdot 2\pi\sqrt{C_f} + \frac{2}{3}\cdot 4C_f\right]\sqrt{\left(\lambda_t x + \frac{2}{9\lambda_t x}\right)^2 + \left(\frac{2}{3}\right)^2}}\right.$$

$$\left.\left[\left(\lambda_t x + \frac{2}{9\lambda_t x}\right)\cdot 2\pi\sqrt{C_f} + \frac{2}{3}\cdot 4C_f\right]\sqrt{\left(\lambda_t x + \frac{2}{9\lambda_t x}\right)^2 + \left(\frac{2}{3}\right)^2}\right\}\mathrm{d}x \quad (6.56)$$

$$= \frac{16}{9}\int_0^1 x\,\mathrm{d}x = \frac{8}{9}$$

It can be deduced that, for the flat airfoil wind turbine that has ideal chord and ideal twist, the thrust coefficient is always 8/9 in stable operation state regardless of the change of design tip speed ratio.

6.3.5 Startup performance

The startup performance of wind turbine plays a significant role in obtaining wider scope of work in order to generate more electricity, and this is particularly true[56] in low velocity zone[57]. The wind turbine is in stable "operation" state even before startup. It must overcome load torque and static friction drag to start, so there is minimum requirement for startup torque. Now let's explore the method of calculating startup torque coefficient.

The magnitude of torque when wind turbine is about to start but not in operation yet is the most significant parameter that represents the performance. The wind turbine will start when the torque overcomes load and friction torque.

The force on wind turbine before it starts is characterized by the following.

(1) As the blades have not been in operation, the inflow angle equals to $\pi/2$ and axial induction speed is 0.

(2) The drag does not generate circumferential component before start and it is only needed to consider the effect of lift.

(3) All blades have large angle of attack, and lift formula of large angle of attack is needed for calculation.

As the drag does not generate circumferential component before wind turbine starts and inflow angle equal to $\pi/2$, the startup torque coefficient is

$$C_M = \frac{B}{\frac{1}{2}\rho U^2 \pi R^3} \int_R dM = \frac{B}{\frac{1}{2}\rho U^2 \pi R^3} \int_R \frac{1}{2}\rho U^2 C C_L r \, dr$$
$$= \frac{B}{\pi} \int_0^1 \left(\frac{C}{R}\right) C_L(x) x \, dx \tag{6.57}$$

Substitute the number of blades, chord and spanwise lift distribution curve into the formula and do integration to obtain the startup torque of wind turbine with any airfoil.

Let's do estimation by taking flat airfoil as an example. According to Formula (6.24) and ideal twist Formula (6.36) of flat airfoil at large angle of attack, the spanwise distribution of lift is

$$C_L(x) = \sin 2\alpha(x) = \sin 2\left[\frac{\pi}{2} - \beta(x)\right] = \sin 2\beta(x)$$
$$= \sin\left(2\arctan\frac{6\lambda_t x}{9\lambda_t^2 x^2 + 2} - 2\arcsin\sqrt{C_f}\right) \quad (6.58)$$

Substitute Formula (6.58) and ideal chord of blade Formula (6.37) into Formula (6.57) to obtain the torque coefficient of ideal wind turbine with flat airfoil.

$$C_M = \frac{B}{\pi}\int_0^1 \left\{ \frac{\frac{8\pi}{9B} \cdot x}{\left[\pi\sqrt{C_f}\left(\lambda_t x + \frac{2}{9\lambda_t x}\right) + \frac{4}{3}C_f\right]\sqrt{\left(\lambda_t x + \frac{2}{9\lambda_t x}\right)^2 + \left(\frac{2}{3}\right)^2}} \right\} C_L(x) x\, dx$$

$$= \frac{8}{9}\int_0^1 \frac{x^2 \sin\left(2\arctan\dfrac{6\lambda_t x}{9\lambda_t^2 x^2 + 2} - 2\arcsin\sqrt{C_f}\right)}{\left[\pi\sqrt{C_f}\left(\lambda_t x + \dfrac{2}{9\lambda_t x}\right) + \dfrac{4}{3}C_f\right]\sqrt{\left(\lambda_t x + \dfrac{2}{9\lambda_t x}\right)^2 + \left(\dfrac{2}{3}\right)^2}}\, dx$$

(6.59)

Set $C_f = 0.01$ and do numerical integration of the above formula to obtain the curve which shows how the torque coefficient changes along the design tip speed ratio in static state, as shown in Figure 6.15. The curve in stable operation state is also included in the figure for easier comparison.

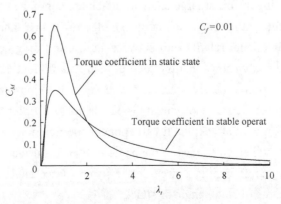

Figure 6.15 Static torque coefficient of ideal wind turbine with flat airfoil

This figure tells us that the startup torque of low speed wind turbine is higher and the static torque coefficient reduces rapidly as design tip speed ratio changes; the static torque coefficient of high speed wind turbine is lower than the torque coefficient in stable operation state.

Calculation of startup torque and startup wind velocity The static startup torque coefficient C_M of wind turbine can be obtained with the approaches described above, and it is assumed they are already given, then the static torque of wind turbine at different wind velocity is

$$M = \frac{1}{2}\rho U^2 \pi R^3 \cdot C_M \qquad (6.60)$$

If the minimum startup torque required by load and friction drag is M_{min}, then the minimum startup velocity is

$$U_{min} = \sqrt{\frac{2M_{min}}{\pi \rho R^3 C_M}} \qquad (6.61)$$

Obviously the startup velocity is inversely proportional to the square root of static torque coefficient.

6.4 Summary of this chapter

This chapter deals with highlighted research on the around flow characteristics of flat airfoil at large angle of attack, and analysis of the experimental data of around flow lift and drag of existing airfoil at large angle of attack are conducted to obtain the total pressure of flat airfoil within Reynolds number range of 10^4-10^6 and an approximate expression of the functional relationship between its lift and drag coefficient and the angle of attack. The study shows that the total pressure coefficient of the flat airfoil at large angle of attack approximately equals to two times of the sinusoidal value of angle of attack and drag coefficient approximately equals to two times of the square of sinusoidal value of angle of attack; the lift component coefficient approximately equals to a sinusoidal value which is 2 times of the angle of attack. Further study reveals that the curve of lift-drag coefficient is a circle in the state of large angle of attack while it is a parabola in the state of small angle of attack.

This chapter also involves the study on the case in which flat airfoil is adopted as the airfoil of wind turbine and the expressions of optimal angle of attack, ideal twist and ideal chord when it is used as ideal blade airfoil are derived; the blade is further adopted as ideal blade to derive the calculation formulas of power, torque, lift and thrust coefficients of the flat airfoil ideal wind turbine running in ideal fluid and actual fluid environment. The research reveals that the sinusoidal value of the optimal angle of attack of flat airfoil equals to the square foot of half of friction drag coefficient of the blade at zero angle of attack; ideal twist is the difference between inflow angle and the optimal angle of attack; ideal chord is a function of the friction drag coefficient related to tip speed ratio at zero angle of attack and the number of blades. The study also shows that the performance of flat airfoil of ideal wind turbine is only related to two parameters—tip speed ratio and friction drag coefficient at zero angle of attack, and is irrelevant to any other factor.

Chapter 7 Function Airfoil and Its Main Performance

The section of blade of wind turbine is airfoil, and airfoil shape has a great influence on performance. However, till today, the functional design of the airfoil has not been achieved, therefore, the analytical calculation for performance of airfoil and wind turbine faces severe challenges. To overcome these challenges and realize blade functional design, this chapter will discuss a method of airfoil profile indicated by analytical function where the change of constant value in the function will bring a cluster of new airfoil, the geometrical significance for function itself and all constant values are absolutely clear and airfoil will be generated in expected direction by adjusting constant values, thereby realizing inverse design. In addition, analytical calculation can be carried out for pressure distribution and lift coefficient of airfoil by function expression, which greatly simplifies design process and significantly improves design efficiency.

7.1 Function construction method for airfoil profile

To carry out analytical calculation for airfoil performance, the function expression of airfoil must be obtained first and definite geometrical significance for parameters in the expression must be given, which is the main task in this section, i.e., studying how to structure airfoil shape by analytical function.

7.1.1 Simplification of Joukowsky airfoil expression

The expression of Joukowsky airfoil profile is[58]

$$z = \sqrt{\frac{1}{4} + \frac{1}{64\varepsilon^2} - y_O^2 - \frac{1}{8\varepsilon} \pm \frac{2\sqrt{3}}{9}\delta(1-2y_O)\sqrt{1-4y_O^2}} \tag{7.1}$$

where, y_O represents non-dimensional x-coordinate of airfoil profile relative to chord (horizontal axis is the same direction as airfoil chord); z represents non-dimensional y-coordinate of airfoil profile relative to chord (wingspan direction is expressed by x); δ represents the ratio of maximum thickness to chord (named as relative thickness); ε

represents the ratio of maximum distance from airfoil mean line to airfoil chord to chord, named as relative curvature. The last term of Formula (7.1) refers to upward profile when it takes plus, and downward profile when it takes minus.

Curvature term is included in radical sign of the formula above, and curvature is at denominator, which is relatively complex. Compared with chord, the curvature is a small quantity, and taylor series expansion is carried out for the first two terms with respect to ε to simplify (the third term contains no small quantity ε), the following codes can be input in Mathematica software:

$$\text{Series}\left[\left(\sqrt{\frac{1}{4}-y_O^2+\frac{1}{64\varepsilon^2}}-\frac{1}{8\varepsilon}\right),\{\varepsilon,0,2\}\right] \quad (7.2)$$

Computational results are

$$(1-4y_O^2)\varepsilon + O[\varepsilon]^3 \quad (7.3)$$

Ignore third order and higher order small quantity in taylor expansion and integrate with the last term of Formula (7.1), which obtains[56]:

$$z = (1-4y_O^2)\varepsilon \pm 0.385\sqrt{1-4y_O^2}(1-2y_O)\delta \quad (7.4)$$

This is the expression of airfoil when chord midpoint is in the origin. Figure 7.1 shows airfoil difference of Formulas (7.1) and (7.4) (when thickness and curvature are 0.2).

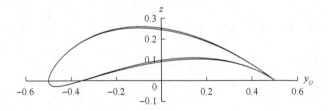

Figure 7.1 Difference between Joukowsky airfoil and simplified function airfoil

Obviously it has little difference before and after function simplification, the figure above shows the situation with large curvature; generally the curvature is much smaller, with little difference in image. The tiny difference will be also further reduced by adjusting constant values.

The first term in Formula (7.4) represents mean line of airfoil, with maximum value of ε, which determines airfoil curvature. The second term represents thickness, which is upward profile if it takes plus, and downward profile when it takes minus. It can be easily proved that maximum distance between upward and downward profiles is δ. $\delta=0.2$, $\varepsilon=0$ and $\delta=0.2$, $\varepsilon=0.1$ are taken respectively, and airfoil shapes in Formula (7.4) are shown in Figures 7.2 and 7.3.

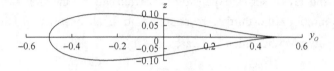

Figure 7.2　Symmetrical airfoil example

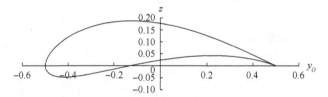

Figure 7.3　Airfoil example (curvature is 0.1)

Moving leading edge to the origin to further simplify Formula (7.4). Taking

$$1 + 2y_O = 2y \tag{7.5}$$

Then

$$1 - 2y_O = 2(1-y) \tag{7.6}$$

Substitute Formulas (7.5) and (7.6) into Formula (7.4):

$$z = 2y \cdot 2(1-y)\varepsilon \pm 0.385\sqrt{2y \cdot 2(1-y)} \cdot 2(1-y)\delta \tag{7.7}$$

Obtain

$$z = 4\varepsilon y(1-y) \pm 1.54\delta y^{0.5}(1-y)^{1.5} \tag{7.8}$$

This is analytical expression of Joukowsky airfoil profile with leading edge at the origin.

7.1.2　Function construction method of general airfoil profile

In the function expression, coefficient and exponent (collectively referred to as constant) have obvious significance on function graph shape. To structure general airfoil profile shape function, coefficient and exponent in simplified Joukowsky airfoil profile expression are extended to general type, by referencing Formula (7.8), airfoil profile function is defined as[59]

$$z = py^a(1-y)^b \pm qy^c(1-y)^d \qquad (7.9)$$

p, a, b, q, c and d in the formula are constants greater than 0. Joukowsky profile function can be considered as a particular case of the formula.

The first term of the formula represents mean line of airfoil, which is controlled by three constants: coefficient p controls height of the whole mean line, a, exponent of y, controls height of front-end mean line, b, exponent of $(1-y)$, controls height of back-end mean line; the second term of the formula represents airfoil thickness whose tendency is controlled by three constants: coefficient q controls the tendency of thickness, c, exponent of y, controls front-end thickness, d, exponent of $(1-y)$ controls back-end thickness.

Increase or decrease of these six constants will have impact on shape, and impact trend relative to datum graph is shown in Table 7.1, here constants that are used to compare with datum graph are $p=0.4$, $a=1$, $b=1$, $q=0.3$, $c=0.5$, $d=1.5$ (Joukowsky airfoil).

Table 7.1 Impact of constant changes on airfoil shape

Constant	Mean line parameters			Thickness parameters		
	p	a	b	q	c	d
Graph when constant increases	Overall rising	Front-end descending	Back-end descending	Overall extension	Front-end narrowing	Back-end narrowing
Datum graph						
Graph when constant decreases	Overall descending	Front-end rising	Back-end rising	Overall narrowing	Front-end extension	Back-end extension

It can be seen in Table 7.1, airfoil profile shape depends on mean line trend and thickness changes. We call this method of structure airfoil profile by adjusting mean line and thickness as "mean line-thickness function structure method".

The impact trend of constant changes on shape shows a strong law. p represents coefficient of mean line, if p increases, airfoil mean line will rise pro rata, and curvature will increase. q represents coefficient of thickness, if q increases, thickness will extend pro rata. The term that the base is y has a great effect on front-end airfoil shape, and the term that base is $(1-y)$ on back-end airfoil shape, both of which are less than 1, therefore, if the exponent increases, the affected term will decrease. Thus it can be seen that geometrical significance of each term, coefficient or exponent in Formula (7.9) is definite, and the expression is not complex (only six constants), therefore, it is easy to structure various shapes of airfoils.

7.1.3 Functional construction method of complex airfoil profile

To structure more complex airfoil shapes, upward and downward profiles may be separated and recombined. Upward and downward profile is shown by subscript u, l, Formula (7.9) can be extended to the following[60]:

$$z_u = p_u y^{a_u} (1-y)^{b_u} + q_u y^{c_u} (1-y)^{d_u} \qquad (7.10)$$

$$z_l = p_l y^{a_l} (1-y)^{b_l} - q_l y^{c_l} (1-y)^{d_l} \qquad (7.11)$$

If the downward profile and mean line always keep datum shape (solid line), the change trend (dotted line) for graph with only upward profile constant is changed is shown in Table 7.2. Correspondingly, if the upward profile and mean line always keep datum shape (solid line), the change trend (dotted line) for graph with only downward profile constant is changed is shown in Table 7.3.

Table 7.2 Impact of upward profile constant changes on airfoil shape

Constant	Mean line parameters of upward profile			Thickness parameters of upward profile		
	p_u	a_u	b_u	q_u	c_u	d_u
Graph when constant increases	Overall rising	Front-end descending	Back-end descending	Overall extension	Front-end narrowing	Back-end narrowing
Graph when constant decreases	Overall descending	Front-end rising	Back-end rising	Overall narrowing	Front-end extension	Back-end extension

Table 7.3 Impact of downward profile constant changes on airfoil shape

Constant	Mean line parameters of downward profile			Thickness parameters of downward profile		
	p_l	a_l	b_l	q_l	c_l	d_l
Graph when constant increases	Overall rising	Front-end descending	Back-end descending	Overall extension	Front-end narrowing	Back-end narrowing
Graph when constant decreases	Overall descending	Front-end rising	Back-end rising	Overall narrowing	Front-end extension	Back-end extension

Above all examples are graph trend with adjusting single constant only based on datum shape. If several constants are adjusted, the change forms for graphs will be diverse, therefore many analytical expressions of the airfoil will be obtained by adjusting constants.

It should be noted that actual mean line and thickness for the airfoil need to be recalculated under upward and downward profile separation and various combinations. Final expression of the mean line is

$$f_\varepsilon = \frac{z_u + z_l}{2}$$
$$= \frac{1}{2}\left[p_u y^{a_u}(1-y)^{b_u} + q_u y^{c_u}(1-y)^{d_u} + p_l y^{a_l}(1-y)^{b_l} - q_l y^{c_l}(1-y)^{d_l}\right] \quad (7.12)$$

Maximum value $f_{\varepsilon\max}$ refers to airfoil curvature. Final distance between upward and downward profile for the airfoil is

$$f_\delta = z_u - z_l$$
$$= p_u y^{a_u}(1-y)^{b_u} + q_u y^{c_u}(1-y)^{d_u} - p_l y^{a_l}(1-y)^{b_l} + q_l y^{c_l}(1-y)^{d_l} \quad (7.13)$$

Maximum value $f_{\delta\max}$ refers to the airfoil thickness, the value can be considered as one of the bound terms in the optimal design.

7.1.4 Functional construction method of smooth trailing edge airfoil profile

Trailing edge for aforesaid airfoil is the intersection point of two curves, and sharp trailing edge airfoil fails to meet actual demand under some conditions, efforts must be made to construct the smooth trailing edge airfoil with analytical function. For example, airfoil at inner side of the blade of the wind turbine always runs under large attack angle, and blunt trailing edge airfoil is often used to improve the lift.

The trailing edge is mainly related to thickness actually, therefore, a thickness term is increased after aforesaid formula to solve the problem. Here the main technique is to apply smooth leading edge method to the trailing edge, and above reference airfoil example analysis is as follows. Reference airfoil profile function is

$$z = 0.4y(1-y) \pm 0.3y^{0.5}(1-y)^{1.5} \quad (7.14)$$

Profile graph indicated in the formula is shown in Figure 7.4(a). Smoothness of the leading edge depends on the exponent of y, 0.5 in thickness term (the second term), the leading edge will be kept smooth when there is slight or no change in such exponent. Smoothness of trailing edge is determined by the exponent of $(1-y)$, when the exponent is adjusted from 1.5 to about 0.5, trailing edge is definitely smooth. However, this adjustment will cause great changes of airfoil profile shape, for example, transposing exponents of both terms, the airfoil will flip horizontally [Figure 7.4(b)]. The solution is to add a thickness term in Formulas (7.9) or (7.14), and transpose exponent position, and decrease coefficient value of the term (to reduce impact on front-end shape), for example, it is transformed to the following form:

$$z = 0.4y(1-y) \pm 0.3y^{0.5}(1-y)^{1.5} \pm 0.1y^{1.5}(1-y)^{0.5} \qquad (7.15)$$

Profile graph indicated in the formula is shown in Figure 7.4(c). If the exponent of y in the third term of Formula (7.15) is increased, i.e., from 1.5 to 6, compared with Figure 7.4(a), the shape of front-end airfoil is basically unchanged, while the trailing edge changes significantly [Figure 7.4(d)].

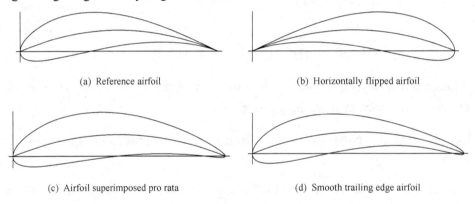

(a) Reference airfoil

(b) Horizontally flipped airfoil

(c) Airfoil superimposed pro rata

(d) Smooth trailing edge airfoil

Figure 7.4 Construction process of smooth trailing edge airfoil

Based on above analysis, the function of smooth trailing edge airfoil profile can be expressed as

$$z = py^a(1-y)^b \pm qy^c(1-y)^d \pm ry^s(1-y)^t \qquad (7.16)$$

where, r, s and t are constants, $r<q$, $s \geq d$, $t \approx c$, and they are constants greater than 0. These three relational expressions can be used as bound terms in smooth trailing edge airfoil optimal design.

Bezier function, similar to the form of above formula, is also appropriate for space curve plotting and widely used in mechanical design field[61], and it is also applied in airfoil design. The expression for n-power Bezier function is

$$P(t) = \sum_{i=0}^{n} B_{i,n} Q_i = \sum_{i=0}^{n} C_n^i t^i (1-t)^{n-i} Q_i \tag{7.17}$$

where, $P(t)$ is any point on the curve, $t \in (0,1)$ is parameter; $B_{i,n}$ is Bernstein primary function; C_n^i is combination number; Q_i is control vertex of Bezier curve.

Structure form for subject variable in Bezier function is similar to the formula in this book, and it is also used for parametric modeling of the airfoil profile[62], with a advantage of benefit to existing airfoil profile approximation, and disadvantage of not being too low for order n, generally exceeding six powers, so the more constant, the larger computation amount; in addition, its coefficient fails to be adjusted arbitrarily, and geometric significance of variables has ambiguity, so it is not good for local shape adjustment of function charting method, but has certain advantage in visual charting.

7.1.5 Function expression of wind turbine blade airfoil

In designing airfoil, an existing airfoil is referenced generally, shape adjustment and performance calculation are made by various methods to determine new airfoil based on optimal performance under constraint condition. There is no exception for wind turbine airfoil design, therefore, it's necessary to express the existing airfoil profile with analytical function.

Airfoil profile is generally described by coordinate database, and it is easy to plot coordinate data to coordinate dot matrix plot, dot matrix is connected orderly to form airfoil image by smooth curve. Approximate airfoil dot matrix by function graph, and airfoil profile integral theory[63] and analytical function linear superposition method[64] are used, the book introduces approximate gradually airfoil dot matrix by adjusting constant in the function based on difference between two graphs. Because geometrical significance of constant value in the function given in the book is definite (Tables 7.2 and 7.3), approximation progress is simple and practicable, generally it takes only several minutes to obtain function expression and graph, which is simple and visual.

Take a airfoil from NACA family airfoil of the US, FFA-W family airfoil of Sweden, DU family airfoil of Netherlands, examples for approximating these three wind turbine airfoils with distinct characteristics are shown in Figures 7.5-7.7, the dot matrix in the figure is original airfoil coordinate dot matrix plot, and the curve is function graph of approximating original airfoil.

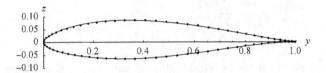

Figure 7.5　Function approximation curve of NACA 63 (2) -215 wind turbine airfoil
(the origin is at leading edge)

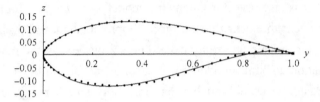

Figure 7.6　Function approximation curve of DU 91-W2-250 wind turbine airfoil
(the origin is at leading edge)

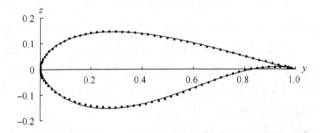

Figure 7.7　Function approximation curve of FFA-W3-301 wind turbine airfoil
(the origin is at leading edge)

Plotting function of Mathematica software can be used to approximate, with the method to show airfoil coordinate data as dot matrix image, and then generate similar airfoil curve by using parameter expression, then superpose two images together, so as to compare their difference, then the airfoil curve is approximated to dot matrix image by adjusting the coefficient and the exponent. After approximation process is completed, airfoil function expression is obtained based on the constant value, that is

NACA 63 (2) -215 airfoil:

$$z_u = 0.2 y^{0.8} (1-y)^{1.25} + 0.11 y^{0.5} (1-y)^{1.5} \qquad (7.18)$$

$$z_l = 0.3 y^{1.1} (1-y)^{5} - 0.33 y^{1.8} (1-y)^{0.66} \qquad (7.19)$$

DU 91-W2-250 airfoil:

$$z_u = 0.2y^{1.1}(1-y)^{2.3} + 0.295y^{0.58}(1-y)^{1.03} \tag{7.20}$$

$$z_l = 0.26y^{0.75}(1-y)^{0.9} - 0.85y^{0.73}(1-y)^{1.6} \tag{7.21}$$

FFA-W3-301 airfoil:

$$z_u = 0.2y^{0.61}(1-y)^{1.6} + 0.24y^{0.48}(1-y)^{1.1} \tag{7.22}$$

$$z_l = 0.27y^{2.6}(1-y)^1 - 0.68y^{0.7}(1-y)^{1.7} \tag{7.23}$$

Similarly, function expressions for other airfoils can be obtained, and adjusting constant value in the expression may adjust the shape slightly so as to design a new airfoil. Constant value range and step length involve in many practical experience, here we provide no more detailed description.

7.1.6 Airfoil parameter expression and airfoil ring view

1) Simple airfoil parameter expression

Set central chord at the origin, it is obtained by Formula (7.4):

$$z = \varepsilon(1+2y)^1(1-2y)^1 \pm 0.385\delta(1+2y)^{\frac{1}{2}}(1-2y)^{\frac{3}{2}} \tag{7.24}$$

The curve indicated in Formula (7.24) can be expressed as parameter equation form as well. Import the parameter θ, make $y=(1/2)\cos\theta$, substitute it into Formula (7.24), obtain

$$z = \varepsilon(1+\cos\theta)^1(1-\cos\theta)^1 \pm 0.385\delta(1+\cos\theta)^{\frac{1}{2}}(1-\cos\theta)^{\frac{3}{2}} \tag{7.25}$$

Parameters θ range in Formula (7.25) is $[0,\pi]$. The formula can be equivalently simplified into single closed curve:

$$z = \varepsilon(1+\cos\theta)^1(1-\cos\theta)^1 + 0.385\delta(1+\cos\theta)^{\frac{1}{2}}(1-\cos\theta)^{\frac{3}{2}} \tag{7.26}$$

or

$$z = \varepsilon(1+\cos\theta)^1(1-\cos\theta)^1 + 0.385\delta\sin\theta(1-\cos\theta)^1 \tag{7.27}$$

Parameters θ range in Formulas (7.26) or (7.27) is $[0, 2\pi]$, it is upward profile when [0, π] is taken, and downward profile when [π, 2π] is taken.

Extending to a more general condition, Formula (7.27) defines single closed parameter expression as an airfoil profile function form:

$$\begin{cases} y = \dfrac{1}{2}\cos\theta \\ z = p(1-\cos\theta)^a (1+\cos\theta)^b + q\sin\theta(1-\cos\theta)^c \end{cases} \quad (7.28)$$

where θ is the parameter; p, a, b and q, d are constants greater than 0, different constants are given to obtain different shapes of airfoil. For example, p is in direct proportion to the curvature, when $p=0$, the airfoil is symmetric airfoil; when $a=b=d=1$, the airfoil is similar to Joukowsky airfoil; when a/b is greater, the airfoil is front arch airfoil; when a/b is smaller, the airfoil is rear arch airfoil; q is in direct proportion to the thickness. Figure 7.8 gives several airfoil profile examples for different constants, and the solid line in airfoil is mean line.

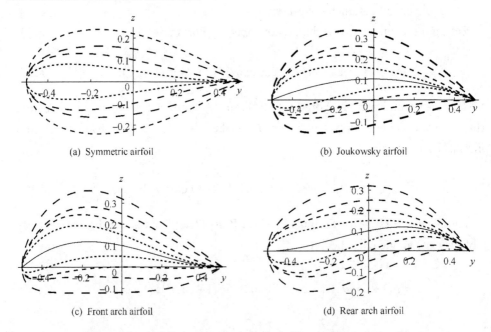

Figure 7.8 Airfoil shape example indicated by parameter expression

The meaning of the parameter θ is as follows (Figure 7.9). Draw circumcircle of the airfoil with the diameter of unit chord, and the azimuth angle of any point P on the circumcircle is θ (the anticlockwise means positive), abscissa of P is $x=(1/2)\cos\theta$.

$P'(x, y)$ on the airfoil corresponds to the nearest point on the airfoil from circumcircle along vertical direction, its x-coordinate is consistent with that of P point, and y-coordinate is the calculated value of airfoil expression.

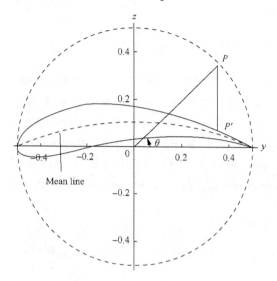

Figure 7.9　Graphical representation for airfoil azimuth

If smooth (not sharp) trailing edge for airfoil is required, one or more thickness terms can be increased, for example, parametric equation for additional one thickness term is

$$z = p(1+\cos\theta)^a (1-\cos\theta)^b + q\sin\theta(1-\cos\theta)^d + r\sin\theta(1+\cos\theta)^f \quad (7.29)$$

r, d are positive constants, $r<q$. Figure 7.10 shows the function shape when $p=2q=0.1$, $r=0.015$, $a=b=d=f=1$.

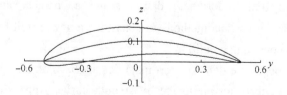

Figure 7.10　Example of airfoil with a certain thickness of trailing edge

2) Parametric expression for complex airfoil

Upward and downward profiles in the airfoil function above discussed have the same constant value, and it is expressed as a single closed parametric expression, i.e., a relatively simple airfoil function.

More generally, upward and downward profiles are expressed respectively to obtain more complex airfoil structure. Another expression that extend Formula (7.26) to the general condition is

$$z = p(1+\cos\theta)^a (1-\cos\theta)^b \pm q(1+\cos\theta)^c (1-\cos\theta)^d \qquad (7.30)$$

Upward and downward profiles in the formula have the same constant value, and to generate more abundant airfoil structures, upward and downward profiles on different airfoils are combined (i.e., different constant values are adopted for upward and downward profiles); to adjust shape or enable to express airfoil shape with smooth trailing edge, one or more thickness terms can be added based on the expression, so a general parametric expression of the airfoil profile is based on these principles:

$$\begin{cases} y = (1/2)\cos\theta \\ z_u = p_u(1+\cos\theta)^{a_u}(1-\cos\theta)^{b_u} + \sum_{i=1}^{M} q_{ui}(1+\cos\theta)^{c_{ui}}(1-\cos\theta)^{d_{ui}} \\ z_l = p_l(1+\cos\theta)^{a_l}(1-\cos\theta)^{b_l} - \sum_{j=1}^{N} q_{lj}(1+\cos\theta)^{c_{lj}}(1-\cos\theta)^{d_{lj}} \end{cases} \qquad (7.31)$$

Subscript u, l represent upward and downward profile of the airfoil; value for p_u, a_u, b_u and q_{ui}, c_{ui}, d_{ui} ($i=1,2,3,\cdots,M$) for the upward airfoil and p_l, a_l, b_l and p_{lj}, a_{lj}, b_{lj} ($j=1,2,3,\cdots,N$) for the downward profile are constants greater than 0, M and N are positive integers.

Taking wind turbine blade airfoil as an example, the function expression for approximating existing airfoil shape is given. In designing airfoil, an existing airfoil is referenced generally (only coordinate data), shape adjustment and performance calculation are made by various methods to determine a new airfoil based on optimal performance under constraint condition.

Plot existing coordinate data to coordinate dot matrix plot, and the dot matrix is connected orderly to form airfoil image by smooth curve. Approximate airfoil dot matrix by function graph, approximating gradually airfoil dot matrix by adjusting the constant in the function based on difference between two graphs.

Take an airfoil from FFA-W family airfoil of Sweden, as shown in Figure 7.11, the dot matrix in the figure is original airfoil coordinate dot matrix plot, and the curve is function graph of approximating original airfoil.

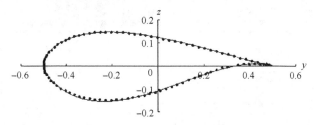

Figure 7.11　Function approximation curve of FFA-W3-301 wind turbine airfoil
(the origin is the center of chord)

After approximation process is completed, the airfoil function expression is obtained based on the constant value.

$$\begin{cases} y = 0.5\cos\theta \\ z_u = 0.043(1+\cos\theta)^{0.61}(1-\cos\theta)^{1.6} + 0.081(1+\cos\theta)^{0.48}(1-\cos\theta)^{1.1} \\ z_l = 0.022(1+\cos\theta)^{2.6}(1-\cos\theta)^{1} - 0.130(1+\cos\theta)^{0.7}(1-\cos\theta)^{1.7} \end{cases} \quad (7.32)$$

Similarly, function expression for other airfoils can be obtained, adjusting constant value in the expression may adjust shape slightly so as to design a new airfoil. The functionalized airfoil can be defined precisely by analytical function, no need to be approximated by coordinate database, which enables to greatly simplify expression manner of the airfoil and facilitates the inverse design.

When the constant for the upward profile is different from that of the downward profile, a single closed parametric formula is replaced with two formulas to express the airfoil, and in this case, it is more simple to express by using general formula (y represents independent variable), particularly moving the origin of coordinates to the leading edge will simplify expression to the full extent.

3) Impact of constant value on airfoil shape

In the function expression, the constant has obvious significance on function graph shape. Azimuth angle θ in the function and value range for function terms are shown in Table 7.4, thus the impact trend of exponent change on the function or the airfoil shape may be estimated. Obviously, $(1+\cos\theta)$ term, between 0 and 1, if on the front end, has great influence on front-end shape, while $(1+\cos\theta)$ term, between 0 and 1, if on the back end, has great influence on back-end shape.

Table 7.4 Value range and characteristic of azimuth angle and function terms

Position	Azimuth angle θ value range	$\cos\theta$ term value range	$(1+\cos\theta)$ term value range and characteristic	$(1-\cos\theta)$ term value range and characteristic
Leading edge point	$\theta=\pi$	$\cos\theta=-1$	0	2
Trailing edge point	$\theta=0$	$\cos\theta=1$	2	0
Midpoint	$\theta=\pm\pi/2$	$\cos\theta=0$	1	1
Front-end	$\pi/2<\theta<3\pi/2$	$-1<\cos\theta<0$	$0<(1+\cos\theta)<1$ The larger exponent, the less value (Significant influence)	$1<(1-\cos\theta)<2$ The larger exponent, the large value (Little influence)
Back-end	$-\pi/2<\theta<\pi/2$	$0<\cos\theta<1$	$1<(1+\cos\theta)<2$ The larger exponent, the large value (Little influence)	$0<(1-\cos\theta)<1$ The larger exponent, the less value (Significant influence)

Taking Formula (7.30) as an example, analyze impact trend of the constant value on shape. The first term of the formula represents the mean line of the airfoil, which is controlled by three constants: coefficient p controls height of the whole mean line (the larger, the higher), exponent a, mainly controls height of front-end mean line (the larger, the lower), exponent b, mainly controls height of back-end mean line (the larger, the lower); the second term of the formula represents airfoil thickness, its trend is controlled by three constants: coefficient q controls the thickness (the larger, the thicker), exponent c, mainly controls front-end thickness (the larger, the thinner), exponent d mainly controls back-end thickness (the larger, the thinner). Impact trend of the constant value on the airfoil shape is shown in Table 7.5.

Table 7.5 Impact trend of constant value change on airfoil shape

Constant change trend	Mean line constant			Thickness constant		
	p	a	b	q	c	d
When constant becomes bigger	Overall rising	Front-end lowers significantly / Back-end lowers slightly	Back-end lowers significantly / Front-end lowers slightly	Overall thickening	Front-end thins significantly / Front-end thins slightly	Back-end thins significantly / Back-end thins slightly
When constant becomes smaller	Overall lowering	Front-end heightens significantly / Front-end heightens slightly	Back-end heightens significantly / Back-end heightens slightly	Overall thinning	Front-end thickens significantly / Front-end thickens slightly	Back-end thickens significantly / Back-end thickens slightly

A simple law can be concluded from Table 7.5: when coefficient becomes larger, mean line will heighten and thicken, with the same change trend; when exponent becomes larger, mean line will lower and thin, with inverse change trend; exponent for $(1+\cos\theta)$ term significantly affects the front end while exponent for $(1-\cos\theta)$ term significantly affects the back end. Knowing this simple law, it is easy to assign value for the constant, thereby generating expected airfoil shape, or promoting approximation to existing airfoil shape.

7.2 Main performance calculation for function airfoil

7.2.1 Function airfoil speed distribution

Make the airfoil function defined by Formula (7.28). With regard to ideal fluid flows at small angle of attack around the airfoil, the airfoil on $h(y,z)$ plane is transformed to the circle on $\zeta(\xi, \eta)$ plane, and circular cylinder formula is used to solve[65].

Complex velocity for circular cylinder with circular rector on $\zeta(\xi, \eta)$ plane is[55]

$$\begin{aligned}\frac{dw}{d\varsigma} = v_\xi - iv_\eta &= U'e^{-i\alpha} - \frac{U'e^{i\alpha}R^2}{\varsigma^2} + \frac{i\Gamma}{2\pi\varsigma} \\ &= U'[\cos\alpha - i\sin\alpha] - U'\frac{(\cos\alpha + i\sin\alpha)R^2}{R^2(\cos 2\theta + i\sin 2\theta)} + U'\frac{i4\pi R\sin\alpha}{2\pi R(\cos\theta + i\sin\theta)} \quad (7.33)\\ &= U'\left[\cos\alpha(1-\cos 2\theta) - \sin\alpha(\sin 2\theta - 2\sin\theta)\right] \\ &\quad - iU'\left[\sin\alpha(1+\cos 2\theta - 2\cos\theta) - \cos\alpha\sin 2\theta\right]\end{aligned}$$

In Formula (7.33), R represents cylinder radius on ζ plane; v_ξ, v_η, U' are velocity component of ξ axis on ζ plane, velocity component of η axis direction and infinite inflow velocity; $\Gamma = 4\pi RU'\sin\alpha$ (the circular rector just moves the rear stagnation point to the trailing edge point). Thus non-dimensional velocity component are respectively.

$$\bar{v}_\xi = \frac{v_\xi}{U'} = \cos\alpha(1-\cos 2\theta) - \sin\alpha(\sin 2\theta - 2\sin\theta) \quad (7.34)$$

$$\bar{v}_\eta = \frac{v_\eta}{U'} = \sin\alpha(1+\cos 2\theta - 2\cos\theta) - \cos\alpha\sin 2\theta \quad (7.35)$$

The literature [66] discussed projection relationship between the shape (not necessarily the airfoil) indicated by single closed function on $h(y, z)$ and the cylinder on plane $\zeta(\xi, \eta)$, and the literature [67] gave further evidence and specified applied condition for transformation relation. On $h(y,z)$ plane, non-dimensional velocity distribution for the shape indicated by single closed function is[63]

$$\bar{v}_y = \frac{v_y}{U} = \frac{\bar{v}_\xi f_1 - \bar{v}_\eta f_2}{f_1^2 + f_2^2} \tag{7.36}$$

$$\bar{v}_z = \frac{v_z}{U} = \frac{\bar{v}_\xi f_2 + \bar{v}_\eta f_1}{f_1^2 + f_2^2} \tag{7.37}$$

In Formulas (7.36) and (7.37)

$$f_1 = \frac{\sin\theta\left[(1-\cos 2\theta) - z'_y \sin 2\theta\right]}{\sin\theta + 2z} \tag{7.38}$$

$$f_2 = \frac{\sin\theta\left[\sin 2\theta + (1-\cos 2\theta)z'_y\right]}{\sin\theta + 2z} \tag{7.39}$$

In regard to function airfoil discussed in the book, to transform above formula and subsequent pressure distribution formula to be derived into the form takes θ as independent variable, for differential of Formula (7.28):

$$dy = -\frac{\sin\theta}{2}d\theta \tag{7.40}$$

$$z'_y = \frac{dz}{dy} = \frac{dz\,d\theta}{d\theta\,dy} = -\frac{2}{\sin\theta}\frac{dz}{d\theta} = -\frac{2}{\sin\theta}z'_\theta \tag{7.41}$$

Substitute Formula (7.40) into Formulas (7.38) and (7.39), obtaining

$$f_1 = \frac{\sin\theta(1-\cos 2\theta) + 2z'_\theta \sin 2\theta}{\sin\theta + 2z} \tag{7.42}$$

$$f_2 = \frac{\sin\theta \sin 2\theta - 2(1-\cos 2\theta)z'_\theta}{\sin\theta + 2z} \tag{7.43}$$

Function airfoil speed square distribution formula is given by Formulas (7.36), (7.37), (7.42) and (7.43)

$$\bar{v}^2 = \bar{v}_y^2 + \bar{v}_z^2 = \frac{\bar{v}_\xi^2 + \bar{v}_\eta^2}{f_1^2 + f_2^2} = \frac{(\sin\theta + 2z)^2\left[\sin(\theta-\alpha) + \sin\alpha\right]^2}{4(z'_\theta)^2 \sin^2\theta + \sin^4\theta} \tag{7.44}$$

7.2.2 Function airfoil pressure distribution

Airfoil pressure distribution is easily to be given by Formula (7.44):

$$C_p = 1 - \left(\bar{v}_y^2 + \bar{v}_z^2\right) = 1 - \frac{(\sin\theta + 2z)^2 \left[\sin(\theta - \alpha) + \sin\alpha\right]^2}{4(z'_\theta)^2 \sin^2\theta + \sin^4\theta} \quad (7.45)$$

It is general formula for pressure distribution that only takes θ as independent variable. Taking the following calculation as an example, set the parametric expression of the airfoil as

$$z = 0.1(1 - \cos\theta)(1 + \cos\theta) + 0.1\sin\theta(1 - \cos\theta) \quad (7.46)$$

Simplify Formula (7.46), and obtain derivatives:

$$z = 0.1\left(\sin^2\theta + \sin\theta - 0.5\sin 2\theta\right) \quad (7.47)$$

$$z'_\theta = 0.1(\sin 2\theta + \cos\theta - \cos 2\theta) \quad (7.48)$$

Substitute into Formula (7.45), obtaining

$$C_p = 1 - \frac{\left(0.2\sin^2\theta + 1.2\sin\theta - 0.1\sin 2\theta\right)^2 \left[\sin(\theta - \alpha) + \sin\alpha\right]^2}{0.04(\sin 2\theta + \cos\theta - \cos 2\theta)^2 \sin^2\theta + \sin^4\theta} \quad (7.49)$$

It can be seen that pressure distribution is expressed as single closed curve function of azimuth angle.

In Formula (7.49), when the angles of attack α are 5° and 10°, ring view for pressure distribution is shown in Figure 7.12, θ is clockwise azimuth angle of the circumcircle from trailing edge.

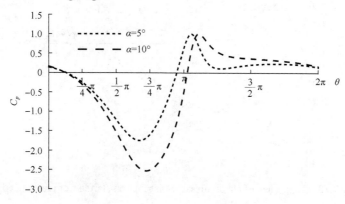

Figure 7.12   Ring view of airfoil pressure distribution

This image reflects the airfoil pressure distribution by taking θ as independent variable, which is called as the ring view of airfoil pressure distribution. Traditionally, the origin of pressure distribution diagram superposes with the leading edge, therefore, azimuth γ started from leading edge point is taken as the independent variable (Figure 7.9), azimuth for upward and downward profiles are represented by subscript u, l, the relationship with θ:

$$\gamma_u = \pi - \theta, \theta \in [0, \pi], \gamma_u \in [0, \pi] \text{ (clockwise as positive)}$$

$$\gamma_l = \theta - \pi, \theta \in [\pi, 2\pi], \gamma_l \in [0, \pi] \text{ (anticlockwise as positive)}$$

Substitute Formula (7.45), the pressure coefficient for upward and downward profiles are respectively:

$$C_{pu} = 1 - \frac{(\sin \gamma_u + 2z)^2 \left[\sin(\gamma_u + \alpha) + \sin \alpha\right]^2}{4\left(z'_\theta \big|_{\theta=\pi-\gamma_u}\right)^2 \sin^2 \gamma_u + \sin^4 \gamma_u} \quad (7.50)$$

$$C_{pl} = 1 - \frac{(\sin \gamma_l - 2z)^2 \left[\sin(\gamma_l - \alpha) - \sin \alpha\right]^2}{4\left(z'_\theta \big|_{\theta=\pi+\gamma_l}\right)^2 \sin^2 \gamma_l + \sin^4 \gamma_l} \quad (7.51)$$

It is general formula for pressure distribution that only takes γ as the independent variable. Substitute Formula (7.47) and derivative Formula (7.48) into the formula above to obtain pressure distribution ring view started from the leading edge, as shown in Figure 7.13.

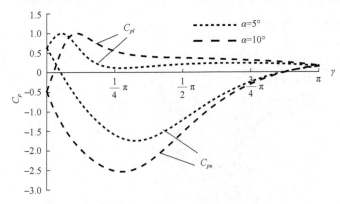

Figure 7.13 Ring view of airfoil pressure distribution started from leading edge

It can be seen from Figures 7.12 and 7.13, the ring view of pressure distribution taking the azimuth as independent variable shows very clearly for leading edge pressure distribution, which has obvious advantages than taking the chord y as independent variable. Figure 7.14 is pressure distribution diagram along chord given by Profili (XFOIL kernel) when Reynolds number $Re=4.3\times10^5$, angle of attack $\alpha=3°$ for a more practical micro-bend airfoil than Joukowsky profile. The expression of the airfoil is

$$z = 0.01\sin^2\theta + 0.05\sin\theta(1-\cos\theta)^{0.6} \tag{7.52}$$

Substitute Formula (7.52) into Formulas (7.50) and (7.51) to obtain ring view of ideal fluid environment pressure distribution, as shown in Figure 7.15.

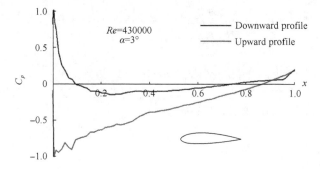

Figure 7.14 Micro-bend airfoil pressure distribution diagram along chord

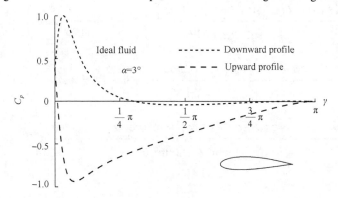

Figure 7.15 Ring view of micro-bend airfoil pressure distribution along azimuth angle

It can be found that by contrast Figures 7.14 and 7.15, pressure distribution has the same trend and approximate pressure value, and the most obvious characteristics is that ring view for pressure distribution at leading edge shows more clear (if thickness is considerable for trailing edge, the same can do it). Reasons caused by error will be described in latter chapters.

7.2.3 Function airfoil lift coefficient calculation

Lift is the most important performance sign of the airfoil, and it can be obtained by direct pressure integration. If airfoil surface pressure is expressed as p, and infinite inflow pressure is expressed as p_0, then the integral of component force that is perpendicular to inflow direction for the force of differential pressure $(p-p_0)$ applied on profile micro-line ds is lift to be solved.

$$dL = -(p-p_0)ds \cdot \sin\left[\left(\arctan z'_y - \frac{\pi}{2}\right) - \alpha\right] = (p-p_0)\cos(\arctan z'_y - \alpha)ds$$

$$= \frac{1}{2}\rho U^2\left[1-\left(\bar{v}_y^2 + \bar{v}_z^2\right)\right]\left[\frac{\cos\alpha}{\sqrt{1+(z'_y)^2}} + \frac{z'_y \sin\alpha}{\sqrt{1+(z'_y)^2}}\right]\sqrt{1+(z'_y)^2}\,dy \quad (7.53)$$

$$= \frac{1}{2}\rho U^2\left[1-\left(\bar{v}_y^2 + \bar{v}_z^2\right)\right](\cos\alpha + z'_y \sin\alpha)dy$$

From Formula (7.44), lift coefficient is

$$C_L = \frac{dL}{\frac{1}{2}\rho U^2} = \oint_S \left[1-\left(\bar{v}_y^2 + \bar{v}_z^2\right)\right](\cos\alpha + z'_y \sin\alpha)dy$$

$$= \int_0^{2\pi}\left\{1 - \frac{(\sin\theta + 2z)^2[\sin(\theta-\alpha) + \sin\alpha]^2}{4(z'_\theta)^2 \sin^2\theta + \sin^4\theta}\right\}\left(\cos\alpha + \frac{-2z'_\theta}{\sin\theta}\sin\alpha\right)\left(-\frac{\sin\theta}{2}d\theta\right)$$

$$= \int_0^{2\pi}\left\{-1 + \frac{(\sin\theta + 2z)^2[\sin(\theta-\alpha) + \sin\alpha]^2}{4(z'_\theta)^2 \sin^2\theta + \sin^4\theta}\right\}\left(\frac{1}{2}\sin\theta\cos\alpha - z'_\theta\sin\alpha\right)d\theta$$

$$= \int_0^{2\pi}\frac{(\sin\theta + 2z)^2[\sin(\theta-\alpha) + \sin\alpha]^2(\sin\theta\cos\alpha - 2z'_\theta\sin\alpha)}{8(z'_\theta)^2\sin^2\theta + 2\sin^4\theta}d\theta$$

(7.54)

This is a general computational formula of lift coefficient. Obviously, after the airfoil function $z=y(\theta)$ is determined, lift coefficient is the function of angle of attack only. Substitute airfoil function and its derivative into it, lift value for different attack angles can be obtained after integration.

To verify correctness of Formula (7.54), Blasius is used to deduce lift coefficient. Set airfoil profile as S, the force along chord direction as Y, the force perpendicular to chord direction as Z, by Blasius formula:

$$Y - iZ = i\frac{1}{2}\rho \oint_S \left(\frac{dw}{dh}\right)^2 dz = i\frac{1}{2}\rho U^2 \oint_S \left(\bar{v}_y - i\bar{v}_z\right)^2 (dy + idz)$$

$$= i\frac{1}{2}\rho U^2 \oint_S \left(\bar{v}_y^2 - \bar{v}_z^2 - i2\bar{v}_y\bar{v}_z\right)(1 + iz_y')dy$$

$$= -\frac{1}{2}\rho U^2 \oint_S \left(\bar{v}_y^2 z_y' - \bar{v}_z^2 z_y' - 2\bar{v}_y\bar{v}_z\right)dy - i\frac{1}{2}\rho U^2 \oint_S \left(-\bar{v}_y^2 + \bar{v}_z^2 - 2\bar{v}_y\bar{v}_z z_y'\right)dy$$

(7.55)

Thus

$$Z = \frac{1}{2}\rho U^2 \oint_S \left(-\bar{v}_y^2 + \bar{v}_z^2 - 2z_y'\bar{v}_y\bar{v}_z\right)dy = \frac{1}{2}\rho U^2 \oint_S \left[-\left(\bar{v}_y^2 + \bar{v}_z^2\right)\right]dy$$

$$= \frac{1}{2}\rho U^2 \int_0^{2\pi} \frac{(\sin\theta + 2z)^2 \left[\sin(\theta - \alpha) + \sin\alpha\right]^2}{4(z_\theta')^2 \sin^2\theta + \sin^4\theta} \frac{\sin\theta}{2} d\theta \qquad (7.56)$$

$$= \frac{1}{2}\rho U^2 \int_0^{2\pi} \frac{(\sin\theta + 2z)^2 \left[\sin(\theta - \alpha) + \sin\alpha\right]^2}{8(z_\theta')^2 \sin\theta + 2\sin^3\theta} d\theta$$

$$Y = -\frac{1}{2}\rho U^2 \oint_S \left(\bar{v}_y^2 z_y' - \bar{v}z^2 z_y' - 2\bar{v}_y\bar{v}_z\right)dy = \frac{1}{2}\rho U^2 \oint_S \left(\bar{v}_y^2 + \bar{v}_z^2\right)z_x' dy$$

(7.57)

$$= \frac{1}{2}\rho U^2 \int_0^{2\pi} \frac{(\sin\theta + 2z)^2 \left[\sin(\theta - \alpha) + \sin\alpha\right]^2}{4(z_\theta')^2 \sin^2\theta + \sin^4\theta} z_\theta' d\theta$$

The lift L that is perpendicular to inflow direction is

$$L = Z\cos\alpha - Y\sin\alpha \qquad (7.58)$$

So lift coefficient is

$$C_L = \frac{1}{\frac{1}{2}\rho U^2}(Z\cos\alpha - Y\sin\alpha)$$

$$= \int_0^{2\pi} \frac{(\sin\theta + 2z)^2 \left[\sin(\theta - \alpha) + \sin\alpha\right]^2 \cos\alpha}{8(z_\theta')^2 \sin\theta + 2\sin^3\theta} d\theta$$

$$- \int_0^{2\pi} \frac{(\sin\theta + 2z)^2 \left[\sin(\theta - \alpha) + \sin\alpha\right]^2 z_\theta' \sin\alpha}{4(z_\theta')^2 \sin^2\theta + \sin^4\theta} d\theta \qquad (7.59)$$

$$= \int_0^{2\pi} \frac{(\sin\theta + 2z)^2 \left[\sin(\theta - \alpha) + \sin\alpha\right]^2 (\sin\theta\cos\alpha - 2z_\theta' \sin\alpha)}{8(z_\theta')^2 \sin^2\theta + 2\sin^4\theta} d\theta$$

The result is the same as Formula (7.54), it is thus clear that the result is the same by two methods.

Substitute aforementioned micro-bend airfoil Formula (7.52) into Formulas (7.54) or (7.59) to obtain circle flow lift value, the result is shown theoretical calculating value curve for ideal fluid in Figure 7.16, which also gave actual fluid that Reynolds number is 4.3×10^5 calculated by Profili software value for comparison.

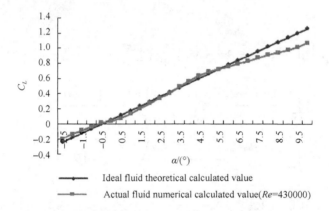

Figure 7.16　Curve comparison between lift coefficient theoretical calculation and numerical calculation

It is seen from Figure 7.16 that theoretical calculating value for lower attack angle is close to calculated value when Reynolds number is 4.3×10^5, lift coefficient for actual fluid trends to reduce when angle of attack is larger, with the reason of viscous interaction: boundary layer displacement thickness for upward and downward profiles is different, which indicates that mean line and trailing edge position for the airfoil are changed, thereby reducing effective attack angle, with more significant deviation for larger attack angle.

Actual fluid flows along the airfoil, and has boundary layer nearby airfoil surface, and external flow velocity for boundary layer is basically calculated based on potential flow theory at present. Internal flow for boundary layer is very complex, which has both laminar flow and turbulent flow, they are calculated by numerical method at present, and analytical method faces huge challenge without analytical solution given, at least not now. A method to be discussed shows approximate solution obtained by empirical formula, however, it takes long time to obtain, so this task is to be discussed in the future.

7.3 Function airfoil and flow calculation application

7.3.1 Airfoil expression extension

It's important to note that pressure distribution Formula (7.45) and lift coefficient Formulas (7.54) or (7.59) are applied widely, but not completely restricted by Formula (7.28) (because z specific expression has not been substituted into the formula), if better airfoil expression will be found, they may be used, provided that single closed condition is met and y expression manner is unchanged (because y expression has been substituted into the formula), therefore, Formula (7.28) can be understood as a airfoil function example that matches the condition.

Here another airfoil function example is given:

$$\begin{cases} y = \frac{1}{2}\cos\theta \\ z = p(1-\cos\theta)^a (1+\cos\theta)^b + q\sin\theta(1-c\cos\theta)(1+d\cos\theta) \end{cases} \quad (7.60)$$

c, d in Formula (7.60) are positive number that is less than 1, c represents the position after adjustment of maximum thickness, d represents the smooth degree after adjustment of trailing edge; Table 7.6 has given eight airfoil shape examples indicated by the function. Obviously this function can easily express the airfoil with smooth trailing edge.

Table 7.6 Airfoil shape example determined by parametric equation

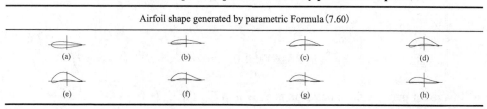

Table 7.7 has given eight pressure distribution examples defined by Formula (7.60), pressure distribution curves at upper and lower part of figure correspond to position of upper and lower airfoil profiles, therefore, coordinate for pressure distribution is expressed by negative pressure coefficient.

Table 7.7 Different function airfoil and pressure distribution examples

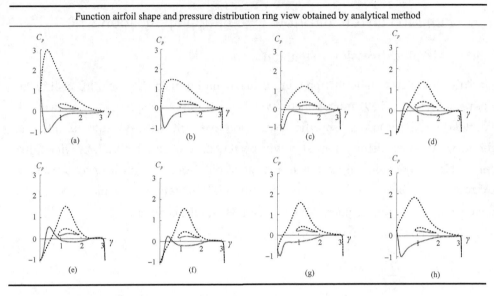

Function airfoil shape and pressure distribution ring view obtained by analytical method

Apparently no matter what manner is used to express the airfoil, pressure distribution and lift are calculated by Formulas (7.45) and (7.54). It is observed that parameters in the airfoil function determine pressure distribution and lift, in this way, one-to-one corresponding function relation is created between parameters and performance.

Parametric expression can also be used to approximate for wind turbine airfoil, here parametric expression approximation curve for DU 91-W2-250 wind turbine airfoil is given:

$$\begin{cases} y = \frac{1}{2}\cos\theta \\ z = 0.011(1-\cos\theta)^{0.6}(1+\cos\theta)^{2.6} + 0.1\sin\theta(1-\cos\theta)^{1.3}(1+\cos\theta)^{0.26} \end{cases} \quad (7.61)$$

Image for parametric expression is shown in Figure 7.17.

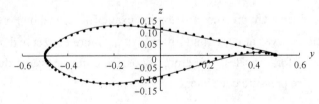

Figure 7.17　Parametric expression approximation curve for DU 91-W2-250 wind turbine airfoil

Above analytical method can be used to calculate speed, pressure distribution and lift coefficient after approximation of wind turbine airfoil by parametric expression, and function graph or the one after fine adjustment is substituted into XFOIL software for numerical calculation and obtain lift, drag and lift-drag ratio.

7.3.2 Main purpose of airfoil analytic calculation

Main purposes of airfoil analytic calculation are:

1) Theoretical calculation

For actual fluid, there is boundary layer nearby the airfoil, but external fluid for boundary layer is considered as ideal fluid, and speed and pressure distribution can be calculated based on potential flow theory; the chapter describes analytic calculation method, explaining function airfoil circle flow analytic method can be used for theoretical calculation.

2) Performance analysis

Analytical calculation method is applicable for flow calculation of ideal fluid, however, it may cause some errors to actual fluid. However, errors do not seriously interfere qualitative analysis for performance, that is to say, change trend for performance is basically consistent, therefore, analytic method can be used for performance estimation and qualitative analysis.

3) Preliminary design for airfoil

Displacement thickness for airfoil boundary layer actually means airfoil shape (although little changes) is changed, so analytical result is only for reference during problem analysis or used as preliminary estimate of performance. Because the airfoil is relative to lift of ideal fluid and lift of actual fluid, in consideration of more simple and quick than numerical calculation, analytic calculation can be used for initial design and analysis, for example, analytic calculation can be used to find an airfoil with better performance, and numerical method is used to calculate actual fluid flow field, thus it can reduce flow field calculation times, improve calculation efficiency and finally is verified by experimental method.

Analytical method is used for initial design or analysis quickly by the following steps:

(1) Give function of approximating the airfoil based on referenced airfoil shape;

(2) Determine change range and step length of each constant in the function (in consideration of constraint condition);

(3) Calculate each change value and lift coefficient of each constant;
(4) Record the maximum lift coefficient and corresponding constant value;
(5) Determine the airfoil based on constant of maximum lift coefficient.

The computer is used to make calculation and comparison for analytical formula, due to mesh generation is not required, the speed ratio is much faster than numerical calculation for fluid field.

Outboard blade for wind turbine has a great influence on power (Section 8.1), airfoil lift-drag ratio is the key factor here, thus it should be design object, and efforts should be made to check drag. However, inner airfoil for wind turbine blade requires maximum lift, so analytical method is more suitable for initial design.

7.3.3 Limitation of airfoil expression

Airfoil expression established in the chapter may express many airfoils by using limited constant, geometrical significance for constant is very definite, thus it is easy to adjust local airfoil shape and whole shape, holding certain advantage in parameterization design. Because the function expression is simple and can be used to generate airfoil image, which allows approximation to existing airfoil to become simple and lay foundation on drawing three-dimensional image of the blade.

However, airfoil expression also has certain limitations. Airfoil function comes from simplification and expansion definition of Joukowsky airfoil, therefore it brings common features of Joukowsky airfoil family inevitably. Joukowsky airfoil family can be obtained by conformal transformation, and studied deeply in potential flow theory, which has very high theoretical value, but seldom applied in engineering practice, with main reason is the airfoil family has thick leading edge and sharp trailing edge; the former meets most airfoil design requirements, but the latter is not allowed for almost all types of fluid machinery. In respect to sharp trailing edge, function structure method for smooth trailing edge airfoil profile has been given, but with the one more thickness term for airfoil function, which becomes complex. Although airfoil expression has some characteristics of Joukowsky airfoil family, the definition is extended to broaden value range of constant, shake off some characteristics of Joukowsky airfoil family, thereby increasing the range of sub-airfoil family.

Another limitation for airfoil expression is that it is only a sub-airfoil family of total airfoil family. In this sub-airfoil family, although infinite number of forms of airfoils can be obtained by change of constant value, it is difficult to express precisely airfoils of another sub-airfoil family, including wind turbine airfoil. For all sub-airfoil families, this problem is inevitable, for example, aforementioned NACA airfoil family of the US, FFA-W airfoil family of Sweden and DU airfoil family of Netherlands have their own series, with obvious characteristic. For over 100 years since generation of airfoils, human being fails to express all airfoils by using one parametric expression, although efforts have been made by multinomial and series method, approximate expression has been only given; however, this expression is difficult to play a role in airfoil design, because constant in the expression has no definite geometrical significance, constant value changes will cause airfoil shape to change unpredictably and wholly, therefore, this airfoil, even "static", is seldom applied.

In short, airfoil expression given in this book can easily and quickly generate precious airfoil profile in its own airfoil family, but generate approximate expression in approximation manner only for airfoils in other sub-airfoil families; degree of approximation will have great difference for different types of airfoils.

7.4 Summary of the chapter

The chapter simplified Joukowsky airfoil profile expression into analytic expression indicated by mean line-thickness function, and further expanded constant range by simple structure characteristic to give the simple method generating different airfoil shapes by adjusting constants, which is applied in airfoil design optimization process. This chapter has also given analytical structure function for complex airfoils, especially the airfoil with smooth trailing edge to avoid many troubles and difficulties of multinomial superposition. Furthermore, this chapter has also given single closed parametric expression for an airfoil profile, which made analytical expression for many airfoil shapes by adjusting constant values with definite geometrical significance, and expressed upward and downward profiles by using the same formula, thereby simplifying subsequent computation process. It is observed that geometrical shape for a complex airfoil can be expressed by analytical function with limited constants which are in a small quantity with definite geometrical significance and easy to use to facilitate the local shape of the airfoil.

This chapter also analyzed theories and engineering application of airfoil expression and analytical calculation method. It also can further expand airfoil expression range, or generate new sub-airfoil family by changing structure form of the expression. In own airfoil family of airfoil expression, precious airfoil can be obtained and calculated by potential flow theory. In preliminary airfoil design, approximate airfoil expression can be obtained by approximating existing airfoil, and a new airfoil is generated by constant assignment, and design process simplification is achieved by resolving preliminary estimate and then more precious numerical calculation or checking performance.

This chapter put forward ring view concept takes the azimuth as independent variable by deducing airfoil pressure distribution formula, effectively resolved unclear pressure distribution diagram for airfoil leading edge. For the given function airfoil, calculation formula with the same lift coefficient is deduced by pressure distribution integral method and Blasius formula which verify each other.

Change trend for constant value in airfoil function may reflect the change trend of shape. This chapter built associative expression between shape and performance, established function relationship of impact of shape on performance, which has important theoretical significance in studying inherent law of airfoil flow.

Chapter 8　Simplified Analysis on Ideal Blade

Chapter 3 gave ideal blade structure based on maximized power. However, it is discussed in theory and difficult to put into practice, because lots of design requirements are proposed for wind turbine actually[68], especially the shape of ideal blade is very complex and difficult to manufacture, its strength failed to meet demand of actual fluid environment. This chapter and later chapters will systematically discuss practicability of ideal blade, that is to say, simplifying blade shape based on ideal blade structure with principal method to approximate curve distribution law of the chord by using straight-line distribution form of the chord to reduce power consumption as far as possible. This chapter will also fully analyze performance when wind turbine runs smoothly composed by simplified blades, so as to illustrate the feasibility and reasonability of this approximation method.

8.1　Blade chord and twist simplification

8.1.1　Simplification purpose and principle

Besides airfoil, blade structure mainly consists of chord and twist. Ideal chord formula is

$$\frac{C}{R} = \frac{16\pi}{9B} \frac{x}{\left[\left(\lambda_t x + \frac{2}{9\lambda_t x}\right)C_L + \frac{2}{3}C_D\right]\sqrt{\left(\lambda_t x + \frac{2}{9\lambda_t x}\right)^2 + \left(\frac{2}{3}\right)^2}} \quad (8.1)$$

If lift coefficient C_L and drag coefficient C_D keep constant value along span, chord curve will become very complex and it is very difficult to manufacture. Readjusting distribution of coefficient C_L and drag coefficient C_D may allow chord curve to become simpler, this means attack angle will be adjusted, i.e., twist must be adjusted to make attack angle change expectedly. However, this step must start from chord curve simplification and be met by backstepping approach.

The main purpose of blade structure simplification is to lower manufacturing difficulty on the premise of reducing power consumption as far as possible. It is easy to manufacture straight-line chord, and the smaller for twist, the easier for manufacturing. The change to straight-line chord from curve chord enables to greatly lower manufacturing difficulty, and twist change has no obvious impact on manufacturing difficulty; obviously chord curve simplification should be given priority, followed by the impact on twist.

In design, generally leading edge and trailing edge for airfoil chord are designed as straight line along span direction, i.e., flat shape for blade is designed as trapezoid, wider at the part that is close to blade root, narrower at the part that is closes to blade tip, this design is benefit to manufacturing and keep performance. Relative chord expression is a straight line after simplification, expressed as

$$\frac{C}{R} = C_r + kx \tag{8.2}$$

C_r and k are constants. Here chord expression is a general formula, parameters C_r and k or specific method should be given. Both parameters are not set arbitrarily, otherwise it may cause deep stall of blade, and may also cause severe decreased efficiency or deviation from design tip speed ratio of maximum efficiency point.

Chord simplification is based on reduce power loss as far as possible, this requires to investigate influence of each blade radial position on power. The following is an example of flat airfoil. According to Formula (6.50), calculation formula for ideal wind turbine power coefficient of flat airfoil is

$$C_P = \frac{16}{9} \int_0^1 \frac{\lambda_1 x^2 \left[\frac{2}{3} \cdot 2\pi \sqrt{C_f} - \left(\lambda_1 x + \frac{2}{9\lambda_1 x} \right) \cdot 4C_f \right]}{\left(\lambda_1 x + \frac{2}{9\lambda_1 x} \right) \cdot 2\pi \sqrt{C_f} + \frac{2}{3} \cdot 4C_f} dx$$

$$= \int_0^1 \frac{16}{9} \frac{\lambda_1 x^2 \left[\frac{\pi}{3} - \left(\lambda_1 x + \frac{2}{9\lambda_1 x} \right) \right] \sqrt{C_f}}{\frac{\pi}{2} \left(\lambda_1 x + \frac{2}{9\lambda_1 x} \right) + \frac{2}{3} \sqrt{C_f}} dx \tag{8.3}$$

Integrand is

$$\frac{\mathrm{d}C_P}{\mathrm{d}x} = \frac{16}{9} \frac{\lambda_t x^2 \left[\frac{\pi}{3} - \left(\lambda_t x + \frac{2}{9\lambda_t x}\right)\sqrt{C_f}\right]}{\frac{\pi}{2}\left(\lambda_t x + \frac{2}{9\lambda_t x}\right) + \frac{2}{3}\sqrt{C_f}} \qquad (8.4)$$

Now observe integrand image, as shown in Figure 8.1.

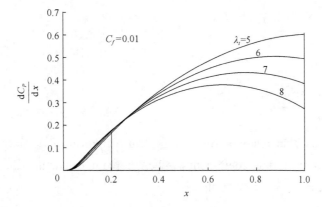

Figure 8.1   Distribution of integrand for power coefficient along span direction

Ideal wind turbine power coefficient is the area under the curve at the designed tip speed ratio. It can be seen from the figure:

(1) External blade has made more contribution to power than internal blade, therefore it is not easy to change chord curve nearby blade tip.

(2) The area (i.e., power coefficient contributed at the section) enclosed by the curve within blade root [0, 0.2] is about 0.015 only, it can be ignored since it is small quantity compared with total area. Therefore, the chord at blade root is not essential, only strength can be reached.

Chord curve should be simplified based on above characteristics.

8.1.2 Influence of drag coefficient on chord

By Formula (3.17), expression for ideal chord is

$$C = \frac{16\pi r}{9B} \frac{1}{\left[\left(\lambda + \frac{2}{9\lambda}\right)C_L + \frac{2}{3}C_D\right]\sqrt{\left(\lambda + \frac{2}{9\lambda}\right)^2 + \left(\frac{2}{3}\right)^2}} \qquad (8.5)$$

Chord expression by ignoring drag coefficient:

$$C' = \frac{16\pi r}{9B} \frac{1}{C_L\left(\lambda + \dfrac{2}{9\lambda}\right)\sqrt{\left(\lambda + \dfrac{2}{9\lambda}\right)^2 + \left(\dfrac{2}{3}\right)^2}} \qquad (8.6)$$

Relative error

$$\frac{C'-C}{C} = \frac{\left(\lambda + \dfrac{2}{9\lambda}\right)C_L + \dfrac{2}{3}C_D}{\left(\lambda + \dfrac{2}{9\lambda}\right)C_L} - 1 = \frac{\dfrac{2}{3}C_D}{\left(\lambda + \dfrac{2}{9\lambda}\right)C_L} \leqslant \frac{\dfrac{2}{3}C_D}{\left(2\sqrt{\lambda}\cdot\sqrt{\dfrac{2}{9\lambda}}\right)C_L} \qquad (8.7)$$

$$= \frac{C_D}{\sqrt{2C_L}} = \frac{1}{\sqrt{2\zeta}} \leqslant \frac{1}{\zeta}$$

It is thus clear that relative error is less than reciprocal of lift-drag ratio, which explains influence from drag coefficient can be ignored in calculating the chord. Thus, the chord curve can be written as follows:

$$\frac{C}{R} = \frac{16\pi}{9B} \frac{x}{C_L\left(\lambda_t x + \dfrac{2}{9\lambda_t x}\right)\sqrt{\left(\lambda_t x + \dfrac{2}{9\lambda_t x}\right)^2 + \left(\dfrac{2}{3}\right)^2}} \qquad (8.8)$$

8.1.3 Simplification method of chord curve

Determining simplified chord formula should be still based on ideal chord. Although the ideal chord curve is complex, the part at blade tip changes slowly, easy to be replaced with straight line. Curvature for blade root changes greatly, however, wind sweeping area at blade root is small, with small contribution for power, so chord at blade root can be expressed by extension line of straight line at blade tip part, with appendant benefit of saving material of blade root and reducing gross weight of blade. This power loss is small, and slight increase of blade length can compensate power loss (but not compensate power coefficient loss).

Replacing ideal chord curve expression with straight line expression is an approximation method, so straight line expression definitely has different forms. Here is a tangent line method.

Now simplify ideal chord curve. In Formula (3.18), drag coefficient is less than lift coefficient by one order of magnitude, so it is ignored; tip speed ratio for high-speed wind turbine is usually large, so reciprocal term is also less than neighboring term by one order of magnitude and it can be ignored, therefore

$$\frac{C}{R} = \frac{16\pi}{9B} \frac{x}{\left[\left(\lambda_t x + \frac{2}{9\lambda_t x}\right) C_L(\alpha_b) + \frac{2}{3} C_D(\alpha_b)\right] \sqrt{\left(\lambda_t x + \frac{2}{9\lambda_t x}\right)^2 + \left(\frac{2}{3}\right)^2}} \quad (8.9)$$

$$\approx \frac{16\pi}{9BC_L(\alpha_b)\lambda_t^2 x}$$

When tip speed ratio is large, the curves before and after the simplification are very close at the blade tip, this simplification meets the principle without changing blade tip chord curve to the greatest extent. A new parameter is defined based on Formula (8.9):

$$y = \frac{CBC_L(\alpha_b)\lambda_t^2}{2\pi R} = \frac{8}{9x} \quad (8.10)$$

The slope of tangent for the parameter curve is

$$k = \frac{dy}{dx} = -\frac{8}{9x^2} \quad (8.11)$$

Specific slope value can be obtained after tangent point is determined. Tangent point should be taken when blade generates the maximum power, and derivative can be computed in Formula (8.9) to find the position where the maximum power occurs. To analyze influence of the smallest blade (the blade using the least material) on performance, tangent point is taken nearby blade tip, set

$$x = \sqrt{8/9} \quad (8.12)$$

Obtain

$$y = \sqrt{8/9}, k = -1 \quad (8.13)$$

Tangential equation passing ($\sqrt{8/9}$, $\sqrt{8/9}$) point nearby blade tip with the slope of -1 is

$$y = 4\sqrt{2}/3 - x \quad (8.14)$$

The corresponding curve and tangent graphs are shown in Figure 8.2.

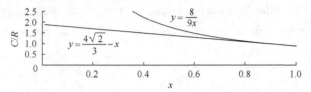

Figure 8.2 Curve-simplified tangent method example

Formula (8.14) is further expressed approximately as

$$\frac{BCC_L(\alpha_b)\lambda_t^2}{2\pi R} = 2 - x \tag{8.15}$$

Transform Formula (8.15) to obtain straight-line approximation formula of relative chord curve.

$$\frac{C}{R} = \frac{2\pi(2-x)}{BC_L(\alpha_b)\lambda_t^2} \tag{8.16}$$

This is a simplified chord formula that is applicable for any airfoil. When the blade number, the lift value for optimal attack angle and design tip speed ratio are given, chord simplification formula will be completely determined. It can be seen from above formula that the chord nearby blade root is two times of that nearby blade tip. The following section will calculate the performance, and examples will show its performance under optimum operating state closely approximates the performance of ideal blade.

Take flat airfoil blade as an example to observe its specific shape. Lift coefficient of optimal attack angle for flat airfoil is

$$C_L(\alpha_b) = 2\pi \sin \alpha_b = 2\pi \sqrt{C_f} \tag{8.17}$$

Substitute Formula (8.17) into Formula (8.16), obtain

$$\frac{C}{R} = \frac{2\pi(2-x)}{BC_L(\alpha_b)\lambda_t^2} = \frac{2-x}{B\lambda_t^2\sqrt{C_f}} \tag{8.18}$$

Set number of blades $B=3$, set tip speed ratio $\lambda_t = 8$, the half of friction drag coefficient $C_f = 0.01$, expression for straight line chord is

$$\frac{C}{R} = \frac{2-x}{B\lambda_t^2 \sqrt{C_f}} = \frac{2-x}{3 \times 8^2 \sqrt{0.01}} = 0.052(2-x) \qquad (8.19)$$

Ideal chord expression that corresponds to this chord formula is

$$\frac{C}{R} = \frac{8\pi}{9B} \frac{x}{\left[\pi\sqrt{C_f}\left(\lambda_t x + \frac{2}{9\lambda_t x}\right) + \frac{4}{3}C_f\right]\sqrt{\left(\lambda_t x + \frac{2}{9\lambda_t x}\right)^2 + \left(\frac{2}{3}\right)^2}}$$

$$= \frac{8\pi}{9 \times 3} \frac{x}{\left[\pi\sqrt{0.01}\left(8x + \frac{2}{9 \times 8x}\right) + \frac{4}{3} \times 0.01\right]\sqrt{\left(8x + \frac{2}{9 \times 8x}\right)^2 + \left(\frac{2}{3}\right)^2}} \qquad (8.20)$$

Distribution diagrams for ideal chord and simplified chord are shown in Figure 8.3.

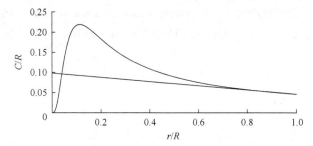

Figure 8.3 Distribution of ideal chord and simplified chord for flat airfoil along span

It is thus clear that simplified chord given in the chapter is the tangent that tangent point is at the inner side of the blade tip.

8.1.4 Blade lift and drag distribution

The chord will no longer conform to blade element-momentum theorem after its change, and blade does not work under design conditions, therefore, it is necessary to readjust lift coefficient distribution along span direction, with the method that lift coefficient conforms to chord formula derived by blade element-momentum theorem, but back calculation for lift coefficient is required, and attack angle is obtained by lift coefficient to calculate twist change. Redistribution expression (influence of lift coefficient is ignored) of lift coefficient is obtained by Formula (8.8):

$$C_L(x) = \frac{16\pi}{9B} \frac{\frac{r}{R}}{\left(\frac{C}{R}\right)\left(\lambda + \frac{2}{9\lambda}\right)\sqrt{\left(\lambda + \frac{2}{9\lambda}\right)^2 + \left(\frac{2}{3}\right)^2}} \quad (8.21)$$

Relative chord C/R can be any expression obtained by using any simplification method. The formula is derived by blade element-momentum theorem (Section 3.4), if lift coefficient produced by new twist or attack angle conforms to the formula, corrected chord and twist conform to blade element-momentum theorem, and will not change design conditions (design tip speed ratio remains unchanged). On the contrary of matching process of solving ideal chord with known twist, here re-distribution of lift coefficient means adjusting lift and attack angle by chord and then adjusting twist, this will match chord with twist to meet design conditions.

Take blade simplification as an example, substitute Formula (8.16) into (8.21) to obtain distribution curve of optimum lift coefficient required under design condition along span direction.

$$C_L(x) = \frac{16\pi}{9B} \frac{\frac{r}{R}}{\left(\frac{C}{R}\right)\left(\lambda + \frac{2}{9\lambda}\right)\sqrt{\left(\lambda + \frac{2}{9\lambda}\right)^2 + \left(\frac{2}{3}\right)^2}}$$

$$= \frac{16\pi}{9B} \frac{x}{\frac{2\pi(2-x)}{BC_L(\alpha_b)\lambda_t^2}\left(\lambda_t x + \frac{2}{9\lambda_t x}\right)\sqrt{\left(\lambda_t x + \frac{2}{9\lambda_t x}\right)^2 + \left(\frac{2}{3}\right)^2}} \quad (8.22)$$

$$= \frac{8}{9} \frac{C_L(\alpha_b)\lambda_t^2 x}{(2-x)\left(\lambda_t x + \frac{2}{9\lambda_t x}\right)\sqrt{\left(\lambda_t x + \frac{2}{9\lambda_t x}\right)^2 + \left(\frac{2}{3}\right)^2}}$$

For flat airfoil

$$C_L(x) = \frac{16\pi}{9} \frac{\sqrt{C_f}\lambda_t^2 x}{(2-x)\left(\lambda_t x + \frac{2}{9\lambda_t x}\right)\sqrt{\left(\lambda_t x + \frac{2}{9\lambda_t x}\right)^2 + \left(\frac{2}{3}\right)^2}} \quad (8.23)$$

Drag distribution is also required when calculating performance. Lift and drag are functions of attack angle, and eliminating attack angle will obtain distribution formula of drag coefficient. For example, in respect to flat airfoil wind turbine blade, drag coefficient distribution function can be obtained by Formulas (6.26), (6.6) and (8.23).

$$C_D(x) = 2C_f + 2\sin^2\alpha(x) = 2C_f + 2\left[\frac{C_L(x)}{2\pi}\right]^2$$

$$= 2C_f + \frac{1}{2\pi^2}\left[\frac{16\pi}{9} \frac{\sqrt{C_f}\lambda_t^2 x}{(2-x)\left(\lambda_t x + \frac{2}{9\lambda_t x}\right)\sqrt{\left(\lambda_t x + \frac{2}{9\lambda_t x}\right)^2 + \left(\frac{2}{3}\right)^2}}\right]^2 \quad (8.24)$$

$$= 2C_f + \frac{128}{81} \frac{C_f \lambda_t^4 x^2}{(2-x)^2\left(\lambda_t x + \frac{2}{9\lambda_t x}\right)^2\left[\left(\lambda_t x + \frac{2}{9\lambda_t x}\right)^2 + \left(\frac{2}{3}\right)^2\right]}$$

8.1.5 Blade attack angle and twist distribution

Attack angle for ideal blade is equal to optimal attack angle, which is a constant. However, attack angle for simplified blade varies from lift distribution. Lift distribution formula is

$$C_L(x) = 2\pi\sin\alpha(x) \quad (8.25)$$

The expression for attack angle distributed along span direction required by Formulas (8.25) and (8.22):

$$\alpha(x) = \arcsin\frac{4C_L(\alpha_b)\lambda_t^2 x}{9\pi(2-x)\left(\lambda_t x + \frac{2}{9\lambda_t x}\right)\sqrt{\left(\lambda_t x + \frac{2}{9\lambda_t x}\right)^2 + \left(\frac{2}{3}\right)^2}} \quad (8.26)$$

In regard to flat airfoil blade, attack angle expression can be obtained by Formulas (8.23) and (8.25):

$$\alpha(x) = \arcsin \frac{8\sqrt{C_f}\lambda_t^2 x}{9(2-x)\left(\lambda_t x + \frac{2}{9\lambda_t x}\right)\sqrt{\left(\lambda_t x + \frac{2}{9\lambda_t x}\right)^2 + \left(\frac{2}{3}\right)^2}} \quad (8.27)$$

Flat airfoil attack angle distribution curve for different tip speed ratio is shown in Figure 8.4.

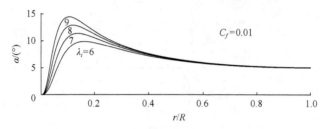

Figure 8.4 Re-distribution curve of attack angle for simplified blade along span

Based on Formulas (3.11) and (8.27), the distribution of twist along span is given:

$$\beta(x) = \varphi(x) - \alpha(x)$$

$$= \arctan \frac{6\lambda_t x}{9\lambda_t^2 x^2 + 2} - \arcsin \frac{4C_L(\alpha_b)\lambda_t^2 x}{9\pi(2-x)\left(\lambda_t x + \frac{2}{9\lambda_t x}\right)\sqrt{\left(\lambda_t x + \frac{2}{9\lambda_t x}\right)^2 + \left(\frac{2}{3}\right)^2}} \quad (8.28)$$

In respect to flat airfoil blade, the distribution of twist along span is

$$\beta(x) = \arctan \frac{6\lambda_t x}{9\lambda_t^2 x^2 + 2} - \arcsin \frac{8\sqrt{C_f}\lambda_t^2 x}{9(2-x)\left(\lambda_t x + \frac{2}{9\lambda_t x}\right)\sqrt{\left(\lambda_t x + \frac{2}{9\lambda_t x}\right)^2 + \left(\frac{2}{3}\right)^2}} \quad (8.29)$$

When $C_f = 0.01$, change curve of twist along tip speed ratio is shown in Figure 8.5. It is observed that twist of simplified blade at blade root is still large, but smaller than ideal blade of which lift coefficient is constant.

Figure 8.6 gave interrelation among inflow angle, corrected attack angle and twist.

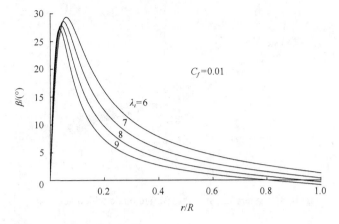

Figure 8.5   Re-distribution curve of twist for simplified blade along span

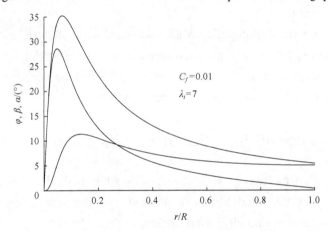

Figure 8.6   Interrelation among inflow angle, attack angle and twist

Since attack angle nearby maximum inflow angle is larger than tangent point at blade tip, which makes maximum twist after simplification become smaller, facilitating the manufacturing of blades.

8.1.6   Outline example for simplified blade

Structure form for simplified blade is determined by blade function, and the function includes three sub-functions: chord function, twist function and airfoil function. An example for simplified blade structure is shown in Figure 8.7, and chord function is defined by Formula (8.19), and twist function is determined by Formula (8.28), and airfoil function is expressed by analytical expression of actual airfoil (Chapter 7).

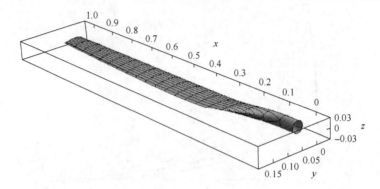

Figure 8.7 Stereogram for simplified blade with blade root

Setting method for blade root and the method generated by function will be discussed in Chapter 12.

The chapter obtained a simplified blade that chord curve is straight line based on ideal blade structure, gave chord and twist formula, and the next chapter will make a comprehensive analysis on performance when the wind turbine operates smoothly to illustrate feasibility and reasonability of the approximation method.

8.2 Performance of simplified blade wind turbine

The previous chapter has given blade simplification method and examples, next the performance for the wind turbine composed by fully new structure blade is calculated, and compared with ideal wind turbine.

8.2.1 Power performance

Substitute simplified blade Chord Formula (8.16) into power coefficient Formula (2.17), and obtain power coefficient.

$$C_P = \frac{B}{\pi}\int_0^1 \lambda_t x \left(\frac{C}{R}\right)\left[\frac{2}{3}C_L - \left(\lambda_t x + \frac{2}{9\lambda_t x}\right)C_D\right]\sqrt{\left(\lambda_t x + \frac{2}{9\lambda_t x}\right)^2 + \left(\frac{2}{3}\right)^2}\, dx$$

$$= \frac{B}{\pi}\int_0^1 \lambda_t x \left[\frac{2\pi(2-x)}{BC_L(\alpha_b)\lambda_t^2}\right]\sqrt{\left(\lambda_t x + \frac{2}{9\lambda_t x}\right)^2 + \left(\frac{2}{3}\right)^2}\left[\frac{2}{3}C_L(x) - \left(\lambda_t x + \frac{2}{9\lambda_t x}\right)C_D(x)\right]dx$$

$$= \int_0^1 \frac{2(2-x)x}{C_L(\alpha_b)\lambda_t}\sqrt{\left(\lambda_t x + \frac{2}{9\lambda_t x}\right)^2 + \left(\frac{2}{3}\right)^2}\left[\frac{2}{3}C_L(x) - \left(\lambda_t x + \frac{2}{9\lambda_t x}\right)C_D(x)\right]dx$$

(8.30)

Chapter 8 Simplified Analysis on Ideal Blade

This is power coefficient formula that is applicable for any airfoil. In the formula, lift coefficient of chord curve expression is lift coefficient that optimal attack angle corresponds to, but other lift and drag coefficient is coefficient re-calculated based on blade element-momentum theory. Taking flat airfoil as an example, substitute lift distribution Formula (8.23), drag distribution Formula (8.24) and optimal attack angle lift Formula (6.34) into above formula to obtain power coefficient.

$$C_P = \int_0^1 \left\{ \frac{2(2-x)x}{2\pi\sqrt{C_f\lambda_t}} \sqrt{\left(\lambda_t x + \frac{2}{9\lambda_t x}\right)^2 + \left(\frac{2}{3}\right)^2} \cdot \frac{2}{3} \cdot \frac{16\pi}{9} \cdot \frac{\sqrt{C_f}\lambda_t^2 x}{(2-x)\left(\lambda_t x + \frac{2}{9\lambda_t x}\right)\sqrt{\left(\lambda_t x + \frac{2}{9\lambda_t x}\right)^2 + \left(\frac{2}{3}\right)^2}} dx \right\}$$

$$- \int_0^1 \frac{2(2-x)x}{2\pi\sqrt{C_f\lambda_t}} \sqrt{\left(\lambda_t x + \frac{2}{9\lambda_t x}\right)^2 + \left(\frac{2}{3}\right)^2} \cdot \left(\lambda_t x + \frac{2}{9\lambda_t x}\right)$$

$$\left[2C_f + \frac{128}{81} \frac{C_f \lambda_t^4 x^2}{(2-x)^2 \left(\lambda_t x + \frac{2}{9\lambda_t x}\right)^2 \left[\left(\lambda_t x + \frac{2}{9\lambda_t x}\right)^2 + \left(\frac{2}{3}\right)^2\right]} \right] dx \quad (8.31)$$

$$= \int_0^1 \left\{ \frac{32}{27} \frac{\lambda_t x^2}{\left(\lambda_t x + \frac{2}{9\lambda_t x}\right)} - \frac{2\sqrt{C_f}}{\pi\lambda_t} x(2-x) \left(\lambda_t x + \frac{2}{9\lambda_t x}\right) \sqrt{\left(\lambda_t x + \frac{2}{9\lambda_t x}\right)^2 + \left(\frac{2}{3}\right)^2} \right.$$

$$\left. \left[1 + \frac{64}{81} \frac{\lambda_t^4 x^2}{(2-x)^2 \left(\lambda_t x + \frac{2}{9\lambda_t x}\right)^2 \left[\left(\lambda_t x + \frac{2}{9\lambda_t x}\right)^2 + \left(\frac{2}{3}\right)^2\right]} \right] \right\} dx$$

The formula is difficult to integrate directly, however design tip speed ratio λ_t is constant, therefore λ_t is assigned with value for numerical integration. To avoid 0 denominator, here numerical integration interval is taken as [0.01, 1], which has little influence on the result (Figure 8.1).

Superimpose dot matrix of numerical integration to power curve of ideal wind turbine for flat airfoil, so as to compare impact of performance between simplified blade and ideal blade structure, as shown in Figure 8.8.

It can be seen from the figure:

(1) Simplified blade has the same performance than ideal blade under the condition that frictional drag coefficient is 0.

(2) With frictional drag, the performance significantly decreases when design tip speed ratio is less than 2, and it slightly decreases when design tip speed ratio is greater than 7, and improves slightly when design tip speed ratio is within 2-7.

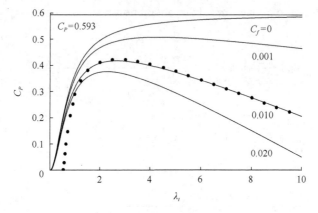

Figure 8.8 Comparison of power coefficient of wind turbine between flat airfoil simplified blade and ideal blade

(3) With frictional drag, all performances have not exceeded limit that is related to tip speed ratio (0 drag line).

(4) Friction drag change is more sensitive than impact of blade structure simplification on performance.

It can be seen that the simplification scheme for blade structure in this book is feasible, and it has no damage property compared with ideal blade when tip speed ratio is greater than 2. Chord curve after simplification is very simple, and is easy to use during manufacturing.

8.2.2 Torque performance

By Formula (2.12), the torque that blade element runs smoothly is

$$dM = r\,dF = \frac{1}{2}\rho U^2 Rr\left(\frac{C}{R}\right)\left[\frac{2}{3}C_L - \left(\lambda + \frac{2}{9\lambda}\right)C_D\right]\sqrt{\left(\lambda + \frac{2}{9\lambda}\right)^2 + \left(\frac{2}{3}\right)^2}\,dr \quad (8.32)$$

For the purpose of integration of above formula, substitute lift and drag coefficients expression redistributed and simplified blade chord Formula (8.16), torque coefficient for simplified blade is obtained

$$C_M = \frac{B}{\pi}\int_0^1 x\left(\frac{C}{R}\right)\left[\frac{2}{3}C_L - \left(\lambda_t x + \frac{2}{9\lambda_t x}\right)C_D\right]\sqrt{\left(\lambda_t x + \frac{2}{9\lambda_t x}\right)^2 + \left(\frac{2}{3}\right)^2}\,dx$$

$$= \frac{B}{\pi}\int_0^1 x\left[\frac{2\pi(2-x)}{BC_L(\alpha_b)\lambda_t^2}\right]\sqrt{\left(\lambda_t x + \frac{2}{9\lambda_t x}\right)^2 + \left(\frac{2}{3}\right)^2}\left[\frac{2}{3}C_L(x) - \left(\lambda_t x + \frac{2}{9\lambda_t x}\right)C_D(x)\right]dx$$

$$= \int_0^1 \frac{2x(2-x)}{C_L(\alpha_b)\lambda_t^2}\sqrt{\left(\lambda_t x + \frac{2}{9\lambda_t x}\right)^2 + \left(\frac{2}{3}\right)^2}\left[\frac{2}{3}C_L(x) - \left(\lambda_t x + \frac{2}{9\lambda_t x}\right)C_D(x)\right]dx$$

(8.33)

This is the torque coefficient formula that is applicable for any airfoil. Taking flat airfoil as an example, substitute lift distribution Formula (8.23), drag distribution Formula (8.24) and optimal attack angle lift Formula (6.34) into Formula (8.33).

$$C_M = \int_0^1 \frac{2x(2-x)}{2\pi\sqrt{C_f}\,\lambda_t^2}\sqrt{\left(\lambda_t x + \frac{2}{9\lambda_t x}\right)^2 + \left(\frac{2}{3}\right)^2}$$

$$\left\{\frac{2}{3}\cdot\frac{16\pi}{9}\frac{\sqrt{C_f}\,\lambda_t^2 x}{(2-x)\left(\lambda_t x + \frac{2}{9\lambda_t x}\right)\sqrt{\left(\lambda_t x + \frac{2}{9\lambda_t x}\right)^2 + \left(\frac{2}{3}\right)^2}}\right.$$

$$\left.-\left(\lambda_t x + \frac{2}{9\lambda_t x}\right)\left[2C_f + \frac{128}{81}\frac{C_f \lambda_t^4 x^2}{(2-x)^2\left(\lambda_t x + \frac{2}{9\lambda_t x}\right)^2\left[\left(\lambda_t x + \frac{2}{9\lambda_t x}\right)^2 + \left(\frac{2}{3}\right)^2\right]}\right]\right\}dx$$

$$= \int_0^1 \left\{\frac{32}{27}\frac{x^2}{\lambda_t x + \frac{2}{9\lambda_t x}} - \frac{2\sqrt{C_f}}{\pi\lambda_t^2}x(2-x)\left(\lambda_t x + \frac{2}{9\lambda_t x}\right)\sqrt{\left(\lambda_t x + \frac{2}{9\lambda_t x}\right)^2 + \left(\frac{2}{3}\right)^2}\right.$$

$$\left.\left[1 + \frac{64}{81}\frac{\lambda_t^4 x^2}{(2-x)^2\left(\lambda_t x + \frac{2}{9\lambda_t x}\right)^2\left[\left(\lambda_t x + \frac{2}{9\lambda_t x}\right)^2 + \left(\frac{2}{3}\right)^2\right]}\right]\right\}dx$$

(8.34)

Now carry out numerical integration for the formula, and superimpose dot matrix of numerical integration to torque coefficient curve of ideal wind turbine for flat airfoil, so as to compare impact of performance between simplified blade and ideal blade, as shown in Figure 8.9.

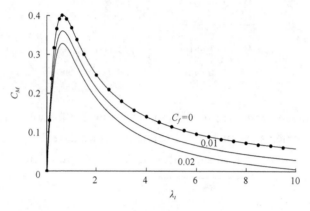

Figure 8.9 Comparison of torque coefficient of wind turbine between flat airfoil simplified blade and ideal blade

It can be seen from the figure, with the drag, performance for simplified blade reduces only when design tip speed ratio is less than 2, and torque performance for simplified blade is basically the same as ideal blade when design tip speed ratio is greater than 2. It can also be seen that all dot matrix is at lower part of the same tip speed ratio 0 drag line, illustrating 0 drag line is performance limit.

8.2.3 Lift performance

Substitute simplified blade chord Formula (8.16) into wind turbine lift coefficient Formula (2.15), and obtain lift coefficient.

$$
\begin{aligned}
C_F &= \frac{B}{\pi}\int_0^1 \left(\frac{C}{R}\right)\left[\frac{2}{3}C_L - \left(\lambda_t x + \frac{2}{9\lambda_t x}\right)C_D\right]\sqrt{\left(\lambda_t x + \frac{2}{9\lambda_t x}\right)^2 + \left(\frac{2}{3}\right)^2}\,dx \\
&= \frac{B}{\pi}\int_0^1 \left[\frac{2\pi(2-x)}{BC_L(\alpha_b)\lambda_t^2}\right]\left[\frac{2}{3}C_L(x) - \left(\lambda_t x + \frac{2}{9\lambda_t x}\right)C_D(x)\right]\sqrt{\left(\lambda_t x + \frac{2}{9\lambda_t x}\right)^2 + \left(\frac{2}{3}\right)^2}\,dx \\
&= \int_0^1 \left[\frac{2(2-x)}{C_L(\alpha_b)\lambda_t^2}\right]\sqrt{\left(\lambda_t x + \frac{2}{9\lambda_t x}\right)^2 + \left(\frac{2}{3}\right)^2}\left[\frac{2}{3}C_L(x) - \left(\lambda_t x + \frac{2}{9\lambda_t x}\right)C_D(x)\right]dx
\end{aligned}
$$

(8.35)

This is the lift coefficient formula that is applicable for any airfoil. Taking flat airfoil as an example, substitute lift distribution Formula (8.23), drag distribution Formula (8.24) and optimal attack angle lift Formula (6.34) into above formula to obtain simplified blade wind turbine lift coefficient.

$$C_F = \int_0^1 \frac{2(2-x)}{2\pi\sqrt{C_f}\lambda_t^2} \sqrt{\left(\lambda_t x + \frac{2}{9\lambda_t x}\right)^2 + \left(\frac{2}{3}\right)^2}$$

$$\left\{\frac{2}{3} \cdot \frac{16\pi}{9} \frac{\sqrt{C_f}\lambda_t^2 x}{(2-x)\left(\lambda_t x + \frac{2}{9\lambda_t x}\right)\sqrt{\left(\lambda_t x + \frac{2}{9\lambda_t x}\right)^2 + \left(\frac{2}{3}\right)^2}}\right.$$

$$\left. -\left(\lambda_t x + \frac{2}{9\lambda_t x}\right)\left[2C_f + \frac{128}{81}\frac{C_f \lambda_t^4 x^2}{(2-x)^2\left(\lambda_t x + \frac{2}{9\lambda_t x}\right)^2\left[\left(\lambda_t x + \frac{2}{9\lambda_t x}\right)^2 + \left(\frac{2}{3}\right)^2\right]}\right]\right\} dx$$

$$= \int_0^1 \left\{\frac{32}{27}\frac{x}{\left(\lambda_t x + \frac{2}{9\lambda_t x}\right)} - \frac{2\sqrt{C_f}}{\pi\lambda_t^2}(2-x)\left(\lambda_t x + \frac{2}{9\lambda_t x}\right)\sqrt{\left(\lambda_t x + \frac{2}{9\lambda_t x}\right)^2 + \left(\frac{2}{3}\right)^2}\right.$$

$$\left.\left[1 + \frac{64}{81}\frac{\lambda_t^4 x^2}{(2-x)^2\left(\lambda_t x + \frac{2}{9\lambda_t x}\right)^2\left[\left(\lambda_t x + \frac{2}{9\lambda_t x}\right)^2 + \left(\frac{2}{3}\right)^2\right]}\right]\right\} dx$$

(8.36)

Now carry out numerical integration for the formula, and superimpose dot matrix of numerical integration to lift coefficient curve of ideal wind turbine for flat airfoil, so as to compare impact of performance between simplified blade and ideal blade, as shown in Figure 8.10.

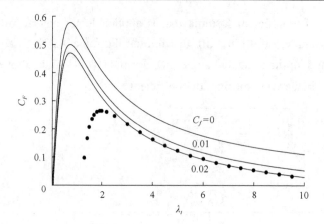

Figure 8.10 Comparison of lift coefficient of wind turbine between flat airfoil simplified blade and ideal blade

It can be seen from the figure, with the drag, performance for simplified blade reduces only when design tip speed ratio is less than 3, and lift performance for simplified blade is basically the same as ideal blade when design tip speed ratio is greater than 3. It also can be seen that all dot matrix is at lower part of the same tip speed ratio 0 drag line, illustrating 0 drag line is lift performance limit.

8.2.4 Thrust performance

Substitute simplified blade chord Formula (8.16) into wind turbine thrust coefficient Formula (2.14), and obtain thrust coefficient.

$$C_T = \frac{B}{\pi}\int_0^1 \left(\frac{C}{R}\right)\sqrt{\left(\lambda+\frac{2}{9\lambda_t x}\right)^2 + \left(\frac{2}{3}\right)^2}\left[\left(\lambda_t x+\frac{2}{9\lambda_t x}\right)C_L + \frac{2}{3}C_D\right]dx$$

$$= \frac{B}{\pi}\int_0^1 \frac{2\pi(2-x)}{BC_L(\alpha_b)\lambda_t^2}\sqrt{\left(\lambda_t x+\frac{2}{9\lambda_t x}\right)^2 + \left(\frac{2}{3}\right)^2}\left[\left(\lambda_t x+\frac{2}{9\lambda_t x}\right)C_L(x) + \frac{2}{3}C_D(x)\right]dx$$

(8.37)

This is the thrust coefficient formula that is applicable for any airfoil. Taking flat airfoil as an example, substitute lift distribution Formula (8.23), drag distribution Formula (8.24) and optimal attack angle lift Formula (6.34) into Formula (8.37) to obtain simplified blade wind turbine thrust coefficient.

$$C_T = \int_0^1 \frac{2(2-x)}{2\pi\sqrt{C_f}\lambda_t^2} \sqrt{\left(\lambda_t x + \frac{2}{9\lambda_t x}\right)^2 + \left(\frac{2}{3}\right)^2} \cdot \left(\lambda_t x + \frac{2}{9\lambda_t x}\right) \cdot \frac{16\pi}{9}$$

$$\frac{\sqrt{C_f}\lambda_t^2 x}{(2-x)\left(\lambda_t x + \frac{2}{9\lambda_t x}\right)\sqrt{\left(\lambda_t x + \frac{2}{9\lambda_t x}\right)^2 + \left(\frac{2}{3}\right)^2}} dx$$

$$= \int_0^1 \left\{ \frac{\int_0^1 \frac{2(2-x)}{2\pi\sqrt{C_f}\lambda_t^2}\sqrt{\left(\lambda_t x + \frac{2}{9\lambda_t x}\right)^2 + \left(\frac{2}{3}\right)^2}\cdot\frac{2}{3}}{\left[2C_f + \frac{128}{81}\cdot\frac{C_f\lambda_t^4 x^2}{(2-x)^2\left(\lambda_t x + \frac{2}{9\lambda_t x}\right)^2\left[\left(\lambda_t x + \frac{2}{9\lambda_t x}\right)^2 + \left(\frac{2}{3}\right)^2\right]}\right]} dx \right\}$$

$$= \int_0^1 \left\{ \frac{\frac{16x}{9} + \frac{4\sqrt{C_f}}{3\pi\lambda_t^2}(2-x)\sqrt{\left(\lambda_t x + \frac{2}{9\lambda_t x}\right)^2 + \left(\frac{2}{3}\right)^2}}{1 + \frac{64}{81}\cdot\frac{\lambda_t^4 x^2}{(2-x)^2\left(\lambda_t x + \frac{2}{9\lambda_t x}\right)^2\left[\left(\lambda_t x + \frac{2}{9\lambda_t x}\right)^2 + \left(\frac{2}{3}\right)^2\right]}} \right\} dx \quad (8.38)$$

Design tip speed ratio λ_t is a constant, therefore λ_t is assigned with value for numerical integration. To avoid 0 denominator, numerical integration interval is taken as [0.01, 1].

Now superimpose dot matrix or dotted line of numerical integration to thrust curve (horizontal straight line in Figure 8.11) of ideal wind turbine for flat airfoil, so as to compare impact of performance between simplified blade and ideal blade, as shown in Figure 8.11.

It can be seen from the figure, simplified blade wind turbine drag coefficient changes has slight impact on thrust coefficient. It also can be seen that when design tip speed ratio is less than 2, it has poor thrust performance for simplified blade. When design tip speed ratio is greater than 2, the thrust coefficient is less than 1 under steady running; when design tip speed ratio is greater than 5, thrust coefficient caused by drag

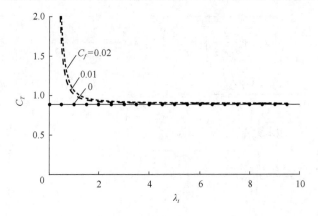

Figure 8.11 Comparison of thrust coefficient of wind turbine between flat airfoil simplified blade and ideal blade

is not more than 0.02 (a little more than the minimum of 8/9), occupying about 2% of total thrust coefficient. These data show that simplified blade still has better wind resistance ability for high speed wind turbine.

8.2.5 Starting performance

Substitute simplified blade chord Formulas (8.16) into (6.57), and obtain torque coefficient.

$$C_M = \frac{B}{\pi}\int_0^1 \left[\frac{2\pi(2-x)}{BC_L(\alpha_b)\lambda_t^2}\right] C_L(x) x\,dx = \int_0^1 \frac{2x(2-x)}{C_L(\alpha_b)\lambda_t^2} C_L(x)\,dx \qquad (8.39)$$

This is starting torque formula of any airfoil wind turbine that is applicable for above simplified program. In regard to flat airfoil, substitute optimal attack angle lift coefficient Formula (6.34), large attack angle lift Formula (6.58) distributed along span and twist Formula (8.29) into Formula (8.39) to obtain starting torque coefficient of simplified blade wind turbine.

$$C_M = \int_0^1 \frac{2x(2-x)}{2\pi\sqrt{C_f}\lambda_t^2}\sin[2\alpha(x)]\,dx = \int_0^1 \frac{x(2-x)}{\pi\sqrt{C_f}\lambda_t^2}\sin[2\beta(x)]\,dx$$

$$= \int_0^1 \frac{x(2-x)}{\pi\sqrt{C_f}\lambda_t^2}\sin\left[2\arctan\frac{6\lambda_t x}{9\lambda_t^2 x^2 + 2} - 2\arcsin\frac{8\sqrt{C_f}\lambda_t^2 x}{9(2-x)\left(\lambda_t x + \frac{2}{9\lambda_t x}\right)\sqrt{\left(\lambda_t x + \frac{2}{9\lambda_t x}\right)^2 + \left(\frac{2}{3}\right)^2}}\right]dx$$

$$(8.40)$$

Set $C_f = 0.01$, carry out numerical integration for Formula (8.40) to obtain the curve of torque coefficient under static condition varying from design tip speed ratio, as shown in Figure 8.12; for ease of comparison, draw the curve of steady running (Figure 8.9) in the figure.

It can be seen from the figure, starting torque for low speed wind turbine is larger, it will increase with design tip speed ratio, and static torque coefficient will decrease quickly; and static torque coefficient for high speed wind turbine is lower than torque coefficient under steady running state.

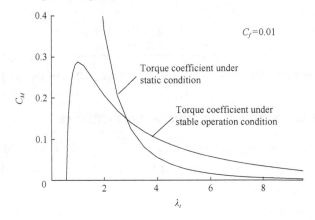

Figure 8.12 Static torque coefficient of flat airfoil simplified blade wind turbine

8.3   Impact analysis of simplification way on performance

8.3.1   Simplification way of blade chord

Keep flat airfoil wind turbine chord distribution curve is straight line, and influence on power is considered by changing chord width ratio. Take simplified chord given in Section 8.1 as base chord, and performance change is studied by magnifying or shrinking the whole blade chord or blade tip and blade root chord and then optimizing distribution of lift, attack angle and twist. Flat shape after chord is changed is shown in Figure 8.13, and the first simplified chord is base chord (chord before changing).

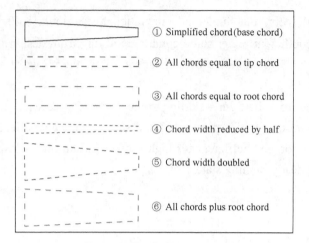

Figure 8.13　Flat shape for simplified chord and blade after changing

8.3.2　Impact of simplification way on performance

Carry out numerical integration for different blades in the figure above, and power coefficient curve for wind turbine composed by relevant blade is shown in Figure 8.14 (corresponding to chord shape number in Figure 8.13).

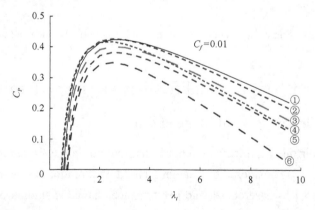

Figure 8.14　Blade power coefficient comparison for corresponding shape in Figure 8.13

It can be seen from the figure, when simplified chord is changed, whether whole or part increase or decrease, it may easily to cause power coefficient decreasing or highest power point deviation, especially for high speed wind turbine, this means chord curve should not be set optionally.

It also can be seen that if chord nearby blade tip remains unchanged, reduce chord at root until equivalent nearby blade tip, and performance reduces little, thereby further saving material. Based on calculation, chord at the root can be adjusted within 1.5-3 times of chord at blade tip, with very low power loss (it is difficult to distinguish difference with chord curve under the scale of above figure). This conclusion is very important to blade design and manufacturing, and it shows to adjust flexibly chord nearby the root provided that the strength is met, and loss performance can be solved by increasing blade length slightly. However, chord at blade tip cannot be changed optionally, because increasing or reducing chord at blade tip will reduce performance significantly. Chord nearby blade tip can be obtained by ideal chord formula.

Examples in the section illustrate the reasonability of ideal chord formula, and it is the basis of chord simplification.

8.4 Summary of this chapter

This chapter analyzed influence degree of each part on power coefficient along span, established chord simplification principle based on chord nearby blade tip and gave a simplification method of making tangent line nearby blade tip. Simplified chord curve obtained by this method is trapezoid, and the blade root is about twice as wide as the blade tip. To allow relationship between simplified chord and twist to meet blade element-momentum theorem, lift distribution along span is re-calculated to obtain the expression of attack angle distribution and new twist distribution. Twist for simplified blade is smaller than the one for ideal blade. Straight chord and smaller twist are beneficial to processing and manufacturing of blades, which is the main purpose of this chapter.

This chapter will make comprehensive analysis for various aerodynamic performances of wind turbine composed of simplified blade, and give integral formula, and carry out detailed calculation for flat airfoil blade and give numerical integration result and graphical representation. The study findings show that when design tip speed ratio is greater than 3, the performance for simplified blade wind turbine is close to ideal wind turbine, it will become poor only when design tip speed ratio is less than 3. Simplified blade is further changed, and the result shows that change of chord nearby blade tip will make performance become poor heavily, but changes at blade root have little influence on performance, which indicates chord at blade tip must be calculated by ideal chord formula and simplified based on it, while chord nearby blade root can be adjusted flexibly based on strength or manufacturing requirements.

Chapter 9 Highest Performance of Practical Wind Turbine

Momentum theory assumes that number of wind turbine blades is infinite, so the air particles flowing through the turbine interact with the blades. But the number of blades of actual wind turbine is finite, so some particles flowing through the turbine interact with the blade, and others will pass through the gap between the blades. The momentum that air transferred to wind rotor under the two conditions is different, or the induction speed is different. If the induction speed at the blade is large, then wind inflow angle will be very small, which will reduce the tangential component of the lift, reduce the power and generate power loss.

Prandtl gave a blade tip loss correction formula. Since blade tip loss has a great impact on blade chord, twist and other performances of the wind turbine, which causes that the highest performance and general performance of ideal wind turbine are far away from actual condition and weakens the guiding function to practical wind turbine design. The maximum performance of practical wind turbine refers to the one with consideration of blade tip loss, but because the blade chord and twist curve is very complex, it must be simplified during manufacturing, and this will further reduce the performance of wind turbine.

Since there are many simplification methods for chord and twist curve, it is not easy to discuss them in details here, the chapter gave the performance of wind turbine without simplification, obviously it is still a unfulfilable performance for human being, but the influence of limited blade numbers is considered to make it more practical, so it is called the highest performance of practical wind turbine here.

9.1 Influence of blade tip loss on structure

9.1.1 Basic expression

For blade tip losses, Prandtl gave an approximate correction method[69], the correction factor improved by Glauert is[22]

$$f = \frac{2}{\pi}\arccos\left\{\exp\left[-\frac{B(1-x)}{2x}\sqrt{1+\frac{\lambda_t^2 x^2}{(1-a)^2}}\right]\right\} \quad (9.1)$$

Where B represents number of blade, a represents axial speed induction factor, and $a=1/3$ under steady running state.

For wind turbine with limited blades, some air particles changed the axial induction factors after interaction with blades; set a_B for axial induction factor of particles at the micro-section of a certain radius on blade, set b_B for tangential induction factor; there are some particles passed through the gap between blades, set $\bar{a}$ for the mean value of axial speed induction factors of all particles in the tiny ring within the same radius, and $\bar{b}$ for the mean value of tangential speed induction factors, and the following is given approximately when tip speed ratio is greater than 3[22]

$$\bar{a} = \frac{1}{3} + \frac{1}{3}f - \frac{1}{3}\sqrt{1-f+f^2} \quad (9.2)$$

$$\bar{b} = \frac{\bar{a}(1-\bar{a}/f)}{\lambda_t^2 x^2} \quad (9.3)$$

The local axial speed induction factor of the particles interacting with the blades is

$$a_B = \bar{a}/f \quad (9.4)$$

$$b_B = \bar{b}/f = \frac{\bar{a}(1-\bar{a}/f)}{\lambda_t^2 x^2 f} \quad (9.5)$$

Curve shape for these parameters is shown in Figure 9.1 (assume $\lambda_t = 7$).

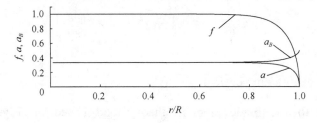

Figure 9.1 The curves of blade tip loss correction factor and average and local axial speed induction factor

As can be seen in the figure, the tip losses occur only in localized parts of the tip; the average axial speed induction factor $\bar{a}$ is reduced to 0 at the tip, but the local axial speed induction factor a_B increases, which will generate huge impacts on inflow angle, lift and other parameters. Obviously the average induction speed factors $\bar{a}$ and $\bar{b}$ are used to calculate lift, thrust, torque and other parameters according to the momentum theory, while the local induction speed factors a_B and b_B are used to calculate according to the blade element theory.

To simplify expressions, $(1-\bar{a}/f)$ is represented by the symbol g, i.e.

$$g = 1 - \bar{a}/f \tag{9.6}$$

The basic formula for follow-up calculation is derived as follows. From Figure 2.2, inflow speed can be obtained based on blade element theory, i.e.

$$v = \sqrt{w^2 + u^2} = \sqrt{(1+b_B)^2 W^2 + (1-a_B)^2 U^2} = \sqrt{(1+\bar{b}/f)^2 \lambda^2 U^2 + (1-\bar{a}/f)^2 U^2}$$

$$= U\sqrt{\left[\lambda + \frac{\bar{a}(1-\bar{a}/f)}{\lambda f}\right]^2 + \left(1 - \frac{\bar{a}}{f}\right)^2} = U\sqrt{\left(\lambda + \frac{\bar{a}g}{\lambda f}\right)^2 + g^2}$$

$$\tag{9.7}$$

Thus the sine and cosine expressions of the inflow angle are

$$\sin\varphi = \frac{u}{v} = \frac{(1-a_B)U}{U\sqrt{\left(\lambda + \frac{\bar{a}g}{\lambda f}\right)^2 + g^2}} = \frac{g}{\sqrt{\left(\lambda + \frac{\bar{a}g}{\lambda f}\right)^2 + g^2}} \tag{9.8}$$

$$\cos\varphi = \frac{w}{v} = \frac{(1+b_B)\lambda U}{U\sqrt{\left(\lambda + \frac{\bar{a}g}{\lambda f}\right)^2 + g^2}} = \frac{\lambda + \frac{\bar{a}g}{\lambda f}}{\sqrt{\left(\lambda + \frac{\bar{a}g}{\lambda f}\right)^2 + g^2}} \tag{9.9}$$

9.1.2 Chord curve correction

According to the blade element theory, and based on Figure 2.2 and Formulas (9.7)-(9.9), the blade element thrust is

$$dT = dL_u + dD_u = dL\cos\varphi + dD\sin\varphi$$
$$= \frac{1}{2}\rho CC_L v^2 \cos\varphi dr + \frac{1}{2}\rho CC_D v^2 \sin\varphi dr$$

$$= \frac{1}{2}\rho U^2 CC_L \left[\left(\lambda + \frac{\overline{ag}}{\lambda f}\right)^2 + g^2\right] \frac{\lambda + \dfrac{\overline{ag}}{\lambda f}}{\sqrt{\left(\lambda + \dfrac{\overline{ag}}{\lambda f}\right)^2 + g^2}} dr$$

$$+ \frac{1}{2}\rho U^2 CC_D \left[\left(\lambda + \frac{\overline{ag}}{\lambda f}\right)^2 + g^2\right] \frac{g}{\sqrt{\left(\lambda + \dfrac{\overline{ag}}{\lambda f}\right)^2 + g^2}} dr \quad (9.10)$$

$$= \frac{1}{2}\rho U^2 C \left[\left(\lambda + \frac{\overline{ag}}{\lambda f}\right) C_L + gC_D\right] \sqrt{\left(\lambda + \frac{\overline{ag}}{\lambda f}\right)^2 + g^2} \, dr$$

According to the momentum theory, the wind thrust to the ring disc of radius r to $r + dr$ is

$$dT = 4\pi\rho U^2 \overline{a}(1-\overline{a})r\,dr \quad (9.11)$$

Set number of blade as B, make the two formulas equal

$$4\pi\rho U^2 \cdot \overline{a}(1-\overline{a})r\,dr = B \cdot \frac{1}{2}\rho U^2 C\left[\left(\lambda + \frac{\overline{ag}}{\lambda f}\right)C_L + gC_D\right]\sqrt{\left(\lambda + \frac{\overline{ag}}{\lambda f}\right)^2 + g^2}\,dr \quad (9.12)$$

Thus the ideal chord formula is

$$\frac{C}{R} = \frac{8\pi}{B}\frac{r}{R} \frac{\overline{a}(1-\overline{a})}{\left[\left(\lambda + \dfrac{\overline{ag}}{\lambda f}\right)C_L + gC_D\right]\sqrt{\left(\lambda + \dfrac{\overline{ag}}{\lambda f}\right)^2 + g^2}}$$

$$= \frac{8\pi}{B} \frac{\overline{a}(1-\overline{a})x}{\left[\left(\lambda_t x + \dfrac{\overline{ag}}{\lambda_t xf}\right)C_L + gC_D\right]\sqrt{\left(\lambda_t x + \dfrac{\overline{ag}}{\lambda_t xf}\right)^2 + g^2}} \quad (9.13)$$

Substitute Formulas (9.1) and (9.2) into Formula (9.13) to obtain specific chord value, and explain with an example as follows. Set optimal attack angle of the airfoil α_b as 3.5°, lift coefficient of the angle is 0.85, drag coefficient is 0.016. Set the airfoil unchanged along the span, and number of blade for wind turbine $B=3$, design tip speed ratio $\lambda_t = 6$, and substitute these parameters into the Formula (9.13), and the chord curve shape of relative chord before and after the correction of blade tip loss is shown in Figure 9.2.

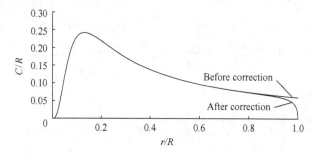

Figure 9.2 Chord curve shape comparison before and after correction of blade tip loss

It can be seen that blade tip loss only affected the blade tip, and the blade tip chord will certainly transit to 0.

9.1.3 Twist curve correction

The corrected formula for inflow angle in reference Formula (2.1) according to the blade element theory is

$$\tan\varphi = \frac{1-a_B}{1+b_B}\frac{1}{\lambda} = \frac{1-\bar{a}/f}{1+\bar{b}/f}\frac{1}{\lambda} = \frac{1-\bar{a}/f}{\lambda + \dfrac{\bar{a}(1-\bar{a}/f)}{\lambda f}} = \frac{g\lambda f}{\lambda^2 f + \bar{a}g} \tag{9.14}$$

In comparison with Formula (3.11), the curve shape before and after the correction of inflow angle is shown in Figure 9.3.

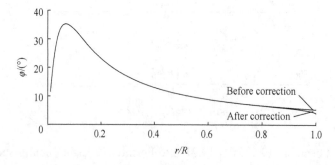

Figure 9.3 Inflow angle curve shape before and after correction of blade tip loss

It can be seen from the figure that inflow angle only has a minor change at blade tip after correction of blade tip loss.

If optimal attack angle of the airfoil α_b is 3.5°, the twist is

$$\beta = \arctan\frac{g\lambda f}{\lambda^2 f + \bar{a}g} - 3.5° \qquad (9.15)$$

An example of the corrected inflow angle and twist distribution curve is shown in Figure 9.4.

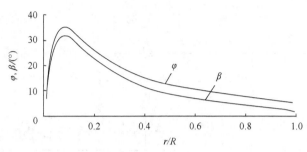

Figure 9.4 Example of inflow angle and twist distribution curve after correction of blade tip loss

9.2 Highest performance calculation for practical wind turbine

9.2.1 Power performance calculation

After consideration of blade tip loss, blade element power is obtained by Formulas (9.7) - (9.9) based on Figure 2.2 as

$$\begin{aligned}
\mathrm{d}P &= \omega r\,\mathrm{d}F = \lambda U(\mathrm{d}L\sin\varphi - \mathrm{d}D\cos\varphi) \\
&= \frac{1}{2}\rho CC_L \lambda Uv^2 \sin\varphi\,\mathrm{d}r - \frac{1}{2}\rho CC_D v^2 \lambda U\cos\varphi\,\mathrm{d}r \\
&= \frac{1}{2}\rho CC_L \lambda U \cdot U^2 \left[\left(\lambda + \frac{\bar{a}g}{\lambda f}\right)^2 + g^2\right] \frac{g}{\sqrt{\left(\lambda + \frac{\bar{a}g}{\lambda f}\right)^2 + g^2}}\,\mathrm{d}r \\
&\quad -\frac{1}{2}\rho CC_D \lambda U \cdot U^2 \left[\left(\lambda + \frac{\bar{a}g}{\lambda f}\right)^2 + g^2\right]\frac{\lambda + \frac{\bar{a}g}{\lambda f}}{\sqrt{\left(\lambda + \frac{\bar{a}g}{\lambda f}\right)^2 + g^2}}\,\mathrm{d}r \\
&= \frac{1}{2}\rho U^3 C\lambda\left[gC_L - \left(\lambda + \frac{\bar{a}g}{\lambda f}\right)C_D\right]\sqrt{\left(\lambda + \frac{\bar{a}g}{\lambda f}\right)^2 + g^2}\,\mathrm{d}r
\end{aligned} \qquad (9.16)$$

By substituting the relative chord Formula (9.13) into Formula (9.16), the power coefficient of wind turbine with B blades is

$$C_P = \frac{B}{\frac{1}{2}\rho U^3 \pi R^2} \int_R \frac{1}{2}\rho U^3 C\lambda \left[gC_L - \left(\lambda + \frac{\bar{a}g}{\lambda f}\right)C_D\right]\sqrt{\left(\lambda + \frac{\bar{a}g}{\lambda f}\right)^2 + g^2}\, dr$$

$$= \frac{B}{\pi}\int_R \left(\frac{C}{R}\right)\cdot\lambda\left[gC_L - \left(\lambda + \frac{\bar{a}g}{\lambda f}\right)C_D\right]\sqrt{\left(\lambda + \frac{\bar{a}g}{\lambda f}\right)^2 + g^2}\, d\left(\frac{r}{R}\right)$$

$$= \frac{B}{\pi}\int_0^1 \frac{8\pi}{B}\frac{\bar{a}(1-\bar{a})x}{\left[\left(\lambda_t x + \frac{\bar{a}g}{\lambda_t xf}\right)C_L + gC_D\right]\sqrt{\left(\lambda + \frac{\bar{a}g}{\lambda f}\right)^2 + g^2}}\cdot \lambda_t x\left[gC_L - \left(\lambda_t x + \frac{\bar{a}g}{\lambda_t xf}\right)C_D\right]$$

$$\sqrt{\left(\lambda + \frac{\bar{a}g}{\lambda f}\right)^2 + g^2}\, dx = \int_0^1 \frac{8\bar{a}(1-\bar{a})\lambda_t x^2\left[gC_L - \left(\lambda_t x + \frac{\bar{a}g}{\lambda_t xf}\right)C_D\right]}{\left(\lambda_t x + \frac{\bar{a}g}{\lambda_t xf}\right)C_L + gC_D}\, dx$$

(9.17)

Set drag coefficient as 0 to obtain the maximum power coefficient that is related to tip speed ratio.

$$C_{P\max} = \int_0^1 \frac{8\bar{a}(1-\bar{a})g\lambda_t x^2}{\lambda_t x + \frac{\bar{a}g}{\lambda_t xf}}\, dx \qquad (9.18)$$

Substitute Formulas (9.1) and (9.2) into Formula (9.18) for numerical integration, the power coefficient change trend after the correction of blade tip loss is shown as dot matrix in Figure 9.5, and the power coefficient curve that the drag is 0 but blade tip loss is not considered has been drawn in the figure.

It can be seen in the figure that limited blade number significantly reduced the power coefficient. It should be noted that chord formula which is not simplified but the blade tip loss has been corrected was used during the derivation process. If simplified chord formula is used, the power coefficient will be further reduced.

If drag is not 0 (lift drag ratio ζ is limited value), it can be obtained by Formula (9.17).

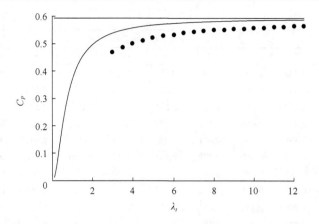

Figure 9.5   Power coefficient change trend after correction of blade tip loss

$$C_P = \int_0^1 \frac{8\bar{a}(1-\bar{a})\lambda_t x^2 \left[gC_L - \left(\lambda_t x + \frac{\bar{a}g}{\lambda_t xf}\right)C_D\right]}{\left(\lambda_t x + \frac{\bar{a}g}{\lambda_t xf}\right)C_L + gC_D} dx \qquad (9.19)$$

$$= \int_0^1 \frac{8\bar{a}(1-\bar{a})\lambda_t x^2 \left[g\zeta - \left(\lambda_t x + \frac{\bar{a}g}{\lambda_t xf}\right)\right]}{\left(\lambda_t x + \frac{\bar{a}g}{\lambda_t xf}\right)\zeta + g} dx$$

The power coefficient obtained from numerical calculation by this integral formula is shown in Table 9.1.

The power loss percentage caused by blade tip loss can be calculated from Tables 5.1 and 9.1, calculation formula: 100× (ideal wind turbine power coefficient−limited blade wind turbine power coefficient)/ideal wind turbine power coefficient, and the result is shown in Table 9.2.

Table 9.1 Power coefficient value after correction of blade tip loss

Lift drag ratio ζ	Design tip speed ratio λ_t									
	1	2	3	4	5	6	7	8	9	10
∞	0.265	0.409	0.472	0.504	0.524	0.537	0.546	0.552	0.557	0.561
1000	0.265	0.408	0.470	0.502	0.521	0.533	0.542	0.548	0.552	0.556
900	0.265	0.408	0.470	0.502	0.521	0.533	0.541	0.547	0.552	0.555
800	0.265	0.408	0.469	0.501	0.520	0.532	0.541	0.547	0.551	0.554
700	0.265	0.408	0.469	0.501	0.520	0.532	0.540	0.546	0.550	0.553
600	0.264	0.407	0.469	0.500	0.519	0.531	0.539	0.545	0.549	0.552

Continued

Lift drag ratio ζ	Design tip speed ratio λ_t									
	1	2	3	4	5	6	7	8	9	10
500	0.264	0.407	0.468	0.499	0.518	0.530	0.538	0.543	0.547	0.550
400	0.264	0.406	0.467	0.498	0.516	0.528	0.535	0.541	0.544	0.547
300	0.264	0.405	0.466	0.496	0.514	0.525	0.532	0.537	0.540	0.542
200	0.263	0.403	0.463	0.492	0.509	0.519	0.525	0.529	0.531	0.532
100	0.260	0.398	0.454	0.481	0.494	0.501	0.504	0.505	0.504	0.502
90	0.260	0.396	0.452	0.478	0.491	0.497	0.500	0.500	0.498	0.496
80	0.259	0.395	0.449	0.475	0.487	0.492	0.494	0.493	0.491	0.488
70	0.258	0.393	0.446	0.470	0.482	0.486	0.487	0.485	0.482	0.477
60	0.257	0.390	0.442	0.465	0.475	0.478	0.477	0.474	0.469	0.463
50	0.255	0.386	0.436	0.457	0.465	0.466	0.463	0.458	0.451	0.444
40	0.252	0.380	0.427	0.445	0.450	0.448	0.443	0.435	0.425	0.414
30	0.248	0.371	0.413	0.426	0.426	0.419	0.409	0.396	0.381	0.365
20	0.240	0.352	0.384	0.387	0.377	0.361	0.340	0.318	0.293	0.268
10	0.215	0.296	0.299	0.273	0.233					

Table 9.2 Power loss percentage (%) caused by blade tip loss

Lift drag ratio ζ	Design tip speed ratio λ_t									
	1	2	3	4	5	6	7	8	9	10
∞	27.89	17.2.49	12.36	9.47	7.77	6.49	5.57	4.92	4.38	4.02
1000	27.84	1	12.36	9.57	7.80	6.45	5.63	4.91	4.47	4.05
900	27.85	17.42	12.40	9.46	7.70	6.52	5.55	5.00	4.40	4.00
800	27.87	17.45	12.45	9.52	7.77	6.44	5.65	4.95	4.37	3.97
700	27.90	17.49	12.34	9.59	7.70	6.56	5.61	4.93	4.37	3.99
600	27.93	17.38	12.42	9.53	7.83	6.54	5.62	4.96	4.42	4.07
500	27.78	17.46	12.37	9.51	7.67	6.58	5.70	4.91	4.40	4.07
400	27.85	17.41	12.37	9.56	7.77	6.57	5.56	4.99	4.36	4.09
300	27.78	17.44	12.49	9.59	7.72	6.59	5.67	5.01	4.46	4.11
200	27.82	17.51	12.38	9.65	7.77	6.65	5.72	5.06	4.51	4.15
100	27.74	17.52	12.57	9.67	7.95	6.64	5.88	5.21	4.65	4.29
90	27.70	17.45	12.44	9.65	7.87	6.68	5.86	5.13	4.69	4.26
80	27.71	17.44	12.58	9.76	7.95	6.72	5.87	5.29	4.64	4.37
70	27.56	17.53	12.51	9.70	8.07	6.86	6.02	5.27	4.81	4.37
60	27.70	17.42	12.65	9.73	8.00	6.87	6.10	5.43	4.86	4.50
50	27.58	17.54	12.60	9.86	8.12	6.98	6.03	5.54	4.97	4.61
40	27.48	17.55	12.79	10.04	8.30	7.16	6.20	5.72	5.15	4.79
30	27.24	17.66	12.92	10.17	8.42	7.28	6.52	6.05	5.48	5.14
20	26.92	17.67	13.17	10.61	9.09	7.97	7.26	6.88	6.65	6.44
10	25.97	17.98	14.17	12.36	11.90					

It can be seen from Table 9.2 that lift-drag ratio has little influence on power loss;

the smaller the design tip speed ratio, the more the power loss.

9.2.2 Torque performance calculation

After considering the blade tip loss, according to Figure 2.2 and Formulas (9.7)-(9.9), the torque of blade element is

$$dM = r\,dF = r(dL\sin\varphi - dD\cos\varphi)$$

$$= \frac{1}{2}\rho CC_L v^2 \sin\varphi \cdot r\,dr - \frac{1}{2}\rho CC_D v^2 \cos\varphi \cdot r\,dr$$

$$= \frac{1}{2}\rho CC_L \cdot U^2 \left[\left(\lambda + \frac{\overline{ag}}{\lambda f}\right)^2 + g^2\right] \frac{g}{\sqrt{\left(\lambda + \frac{\overline{ag}}{\lambda f}\right)^2 + g^2}} r\,dr \qquad (9.20)$$

$$-\frac{1}{2}\rho CC_D \cdot U^2 \left[\left(\lambda + \frac{\overline{ag}}{\lambda f}\right)^2 + g^2\right] \frac{\lambda + \frac{\overline{ag}}{\lambda f}}{\sqrt{\left(\lambda + \frac{\overline{ag}}{\lambda f}\right)^2 + g^2}} r\,dr$$

$$= \frac{1}{2}\rho U^2 C \left[gC_L - \left(\lambda + \frac{\overline{ag}}{\lambda f}\right)C_D\right]\sqrt{\left(\lambda + \frac{\overline{ag}}{\lambda f}\right)^2 + g^2} \cdot r\,dr$$

By relative chord Formula (9.13), the torque coefficient of wind turbine with B blades is

$$C_M = \frac{B}{\frac{1}{2}\rho U^2 \pi R^3} \int_{R2}^{} \frac{1}{2}\rho U^2 C\left[gC_L - \left(\lambda + \frac{\overline{ag}}{\lambda f}\right)C_D\right]\sqrt{\left(\lambda + \frac{\overline{ag}}{\lambda f}\right)^2 + g^2} \cdot r\,dr$$

$$= \frac{B}{\pi}\int_R \left(\frac{C}{R}\right)\cdot\left[gC_L - \left(\lambda + \frac{\overline{ag}}{\lambda f}\right)C_D\right]\sqrt{\left(\lambda + \frac{\overline{ag}}{\lambda f}\right)^2 + g^2} \cdot \left(\frac{r}{R}\right)d\left(\frac{r}{R}\right)$$

$$= \frac{B}{\pi}\int_0^1 \frac{8\pi}{B}\cdot\frac{\overline{a}(1-\overline{a})x}{\left[\left(\lambda_t x + \frac{\overline{ag}}{\lambda_t xf}\right)C_L + gC_D\right]\sqrt{\left(\lambda + \frac{\overline{ag}}{\lambda f}\right)^2 + g^2}}\cdot\left[gC_L - \left(\lambda_t x + \frac{\overline{ag}}{\lambda_t xf}\right)C_D\right] \qquad (9.21)$$

$$\sqrt{\left(\lambda + \frac{\overline{ag}}{\lambda f}\right)^2 + g^2}\cdot x\,dx = \int_0^1 \frac{8\overline{a}(1-\overline{a})x^2\left[gC_L - \left(\lambda_t x + \frac{\overline{ag}}{\lambda_t xf}\right)C_D\right]}{\left(\lambda_t x + \frac{\overline{ag}}{\lambda_t xf}\right)C_L + gC_D}dx$$

Let drag coefficient equals to 0, the maximum torque coefficient associated with tip speed ratio could be obtained

$$C_{M\max} = \int_0^1 \frac{8g\bar{a}(1-\bar{a})x^2}{\lambda_t x + \dfrac{ag}{\lambda_t xf}} dx \qquad (9.22)$$

Substitute the Formulas (9.1) and (9.2) into Formula (9.22) for numerical integration, the torque coefficient change trend after correction of blade tip loss is shown as dot matrix in Figure 9.6, and the torque coefficient curve that the drag is 0 but blade tip loss is not considered has been drawn in the figure.

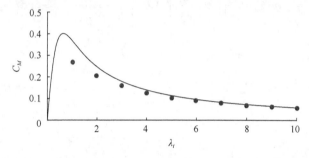

Figure 9.6 Torque coefficient change trend after correction of blade tip loss

It can be seen in the figure that when tip speed ratio is smaller, the limited blade number significantly reduces the torque coefficient. When lift drag ratio is infinitely great, the torque coefficient loss is shown in Table 9.3.

Table 9.3 Influence of blade tip loss on torque coefficient

Tip speed ratio λ_t	Theoretical value for torque coefficient		Torque loss percentage /%
	Without consideration of tip loss	After consideration of tip loss	
1	0.368	0.265	27.91
2	0.248	0.205	17.43
3	0.179	0.157	12.37
4	0.139	0.126	9.52
5	0.114	0.105	7.71
6	0.096	0.089	6.48
7	0.083	0.078	5.58
8	0.073	0.069	4.91
9	0.065	0.062	4.38
10	0.058	0.056	3.95

If the drag is not 0 (lift drag ratio ζ is limited value), it can be obtained by

Formula (9.21):

$$C_M = \int_0^1 \frac{8\bar{a}(1-\bar{a})x^2\left[gC_L - \left(\lambda_t x + \frac{\bar{a}g}{\lambda_t xf}\right)C_D\right]}{\left(\lambda_t x + \frac{\bar{a}g}{\lambda_t xf}\right)C_L + gC_D} dx = \int_0^1 \frac{8\bar{a}(1-\bar{a})x^2\left[g\zeta - \left(\lambda_t x + \frac{\bar{a}g}{\lambda_t xf}\right)\right]}{\left(\lambda_t x + \frac{\bar{a}g}{\lambda_t xf}\right)\zeta + g} dx$$

(9.23)

The torque coefficient obtained from numerical calculation by this integral formula is shown in Table 9.4.

Table 9.4 Torque coefficient value after correction of blade tip loss

Lift drag ratio ζ	Design tip speed ratio λ_t									
	1	2	3	4	5	6	7	8	9	10
∞	0.265	0.205	0.157	0.126	0.105	0.089	0.078	0.069	0.062	0.056
1000	0.265	0.204	0.157	0.125	0.104	0.089	0.077	0.068	0.061	0.056
900	0.265	0.204	0.157	0.125	0.104	0.089	0.077	0.068	0.061	0.056
800	0.265	0.204	0.156	0.125	0.104	0.089	0.077	0.068	0.061	0.055
700	0.264	0.204	0.156	0.125	0.104	0.089	0.077	0.068	0.061	0.055
600	0.264	0.204	0.156	0.125	0.104	0.089	0.077	0.068	0.061	0.055
500	0.264	0.203	0.156	0.125	0.104	0.088	0.077	0.068	0.061	0.055
400	0.264	0.203	0.156	0.125	0.103	0.088	0.077	0.068	0.061	0.055
300	0.263	0.203	0.155	0.124	0.103	0.088	0.076	0.067	0.060	0.054
200	0.262	0.202	0.154	0.123	0.102	0.087	0.075	0.066	0.059	0.053
100	0.259	0.199	0.151	0.120	0.099	0.084	0.072	0.063	0.056	0.050
90	0.259	0.198	0.151	0.120	0.098	0.083	0.072	0.063	0.056	0.050
80	0.258	0.197	0.150	0.119	0.098	0.082	0.071	0.062	0.055	0.049
70	0.257	0.196	0.149	0.118	0.097	0.081	0.070	0.061	0.054	0.048
60	0.255	0.195	0.148	0.117	0.095	0.080	0.068	0.060	0.052	0.047
50	0.253	0.193	0.146	0.115	0.093	0.078	0.067	0.058	0.050	0.045
40	0.250	0.190	0.143	0.112	0.091	0.075	0.064	0.055	0.048	0.042
30	0.245	0.185	0.138	0.107	0.086	0.070	0.059	0.050	0.043	0.037
20	0.235	0.176	0.129	0.098	0.076	0.061	0.050	0.041	0.033	0.027
10	0.207	0.148	0.101	0.070	0.049	0.033	0.021	0.012	0.005	

9.2.3 Lift performance calculation

After consideration of blade tip loss, blade element lift is obtained by Formulas

(9.7) - (9.9) based on Figure 2.2[69]

$$\begin{aligned}
\mathrm{d}F &= \mathrm{d}L\sin\varphi - \mathrm{d}D\cos\varphi = \frac{1}{2}\rho C C_L v^2 \sin\varphi \mathrm{d}r - \frac{1}{2}\rho C C_D v^2 \cos\varphi \mathrm{d}r \\
&= \frac{1}{2}\rho C C_L \cdot U^2 \left[\left(\lambda + \frac{\bar{a}g}{\lambda f}\right)^2 + g^2\right] \frac{g}{\sqrt{\left(\lambda + \frac{\bar{a}g}{\lambda f}\right)^2 + g^2}} \mathrm{d}r \\
&\quad - \frac{1}{2}\rho C C_D \cdot U^2 \left[\left(\lambda + \frac{\bar{a}g}{\lambda f}\right)^2 + g^2\right] \frac{\lambda + \frac{\bar{a}g}{\lambda f}}{\sqrt{\left(\lambda + \frac{\bar{a}g}{\lambda f}\right)^2 + g^2}} \mathrm{d}r \\
&= \frac{1}{2}\rho U^2 C \left[gC_L - \left(\lambda + \frac{\bar{a}g}{\lambda f}\right)C_D\right]\sqrt{\left(\lambda + \frac{\bar{a}g}{\lambda f}\right)^2 + g^2} \, \mathrm{d}r
\end{aligned} \quad (9.24)$$

In respect to wind turbine composed of B blades, substitute relative chord Formula (9.13) into Formula (9.24) for integration to obtain lift coefficient.

$$\begin{aligned}
C_F &= \frac{B}{\frac{1}{2}\rho U^2 \pi R^2} \int_R \frac{1}{2}\rho U^2 C \left[gC_L - \left(\lambda + \frac{\bar{a}g}{\lambda f}\right)C_D\right]\sqrt{\left(\lambda + \frac{\bar{a}g}{\lambda f}\right)^2 + g^2} \, \mathrm{d}r \\
&= \frac{B}{\pi}\int_R \left(\frac{C}{R}\right)\cdot\left[gC_L - \left(\lambda + \frac{\bar{a}g}{\lambda f}\right)C_D\right]\sqrt{\left(\lambda + \frac{\bar{a}g}{\lambda f}\right)^2 + g^2} \, \mathrm{d}\left(\frac{r}{R}\right) \\
&= \frac{B}{\pi}\int_0^1 \frac{8\pi}{B} \frac{\bar{a}(1-\bar{a})x}{\left[\left(\lambda_1 x + \frac{\bar{a}g}{\lambda_1 xf}\right)C_L + gC_D\right]\sqrt{\left(\lambda + \frac{\bar{a}g}{\lambda f}\right)^2 + g^2}} \cdot \\
&\quad \left[gC_L - \left(\lambda_1 x + \frac{\bar{a}g}{\lambda_1 xf}\right)C_D\right]\sqrt{\left(\lambda + \frac{\bar{a}g}{\lambda f}\right)^2 + g^2} \, \mathrm{d}x \\
&= \int_0^1 \frac{8\bar{a}(1-\bar{a})x\left[gC_L - \left(\lambda_1 x + \frac{\bar{a}g}{\lambda_1 xf}\right)C_D\right]}{\left(\lambda_1 x + \frac{\bar{a}g}{\lambda_1 xf}\right)C_L + gC_D} \, \mathrm{d}x
\end{aligned} \quad (9.25)$$

Set drag coefficient as 0 to obtain the maximum lift coefficient that is related to tip speed ratio

$$C_{M\max} = \int_0^1 \frac{8g\bar{a}(1-\bar{a})x}{\lambda_t x + \dfrac{\overline{ag}}{\lambda_t x f}} dx \qquad (9.26)$$

Substitute Formulas (9.1) and (9.2) into Formula (9.26) for numerical integration, the lift coefficient change trend after correction of blade tip loss is shown as dot matrix in Figure 9.7, and lift coefficient curve that the drag is 0 but blade tip loss is not considered has been drawn in the figure.

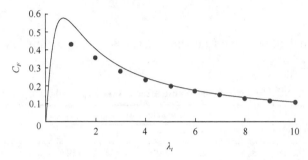

Figure 9.7 Lift coefficient change trend after correction of blade tip loss

It can be seen in the Figure 9.7 that when tip speed ratio is small, the limited blade number significantly reduces the lift coefficient. When lift drag ratio is infinitely great, lift coefficient loss is shown in Table 9.5.

Table 9.5 Influence of blade tip loss on lift coefficient

Tip speed ratio λ_t	Theoretical value for lift coefficient		Lift loss percentage /%
	Without consideration of tip loss	After consideration of tip loss	
1	0.554	0.432	22.02
2	0.406	0.356	12.32
3	0.307	0.282	8.14
4	0.246	0.231	6.10
5	0.204	0.195	4.41
6	0.174	0.168	3.45
7	0.152	0.147	3.29
8	0.135	0.131	2.96
9	0.121	0.118	2.48
10	0.110	0.108	1.82

If drag is not 0 (lift drag ratio ζ is limited value), it can be obtained by Formula (9.25).

$$C_F = \int_0^1 \frac{8\bar{a}(1-\bar{a})x\left[gC_L - \left(\lambda_t x + \frac{\overline{ag}}{\lambda_t xf}\right)C_D\right]}{\left(\lambda_t x + \frac{\overline{ag}}{\lambda_t xf}\right)C_L + gC_D} dx \qquad (9.27)$$

$$= \int_0^1 \frac{8\bar{a}(1-\bar{a})x\left[g\zeta - \left(\lambda_t x + \frac{\overline{ag}}{\lambda_t xf}\right)\right]}{\left(\lambda_t x + \frac{\overline{ag}}{\lambda_t xf}\right)\zeta + g} dx$$

The lift coefficient obtained from numerical calculation by this integral formula is shown in Table 9.6.

Table 9.6 Lift coefficient value after correction of blade tip loss

Lift drag ratio ζ	Design tip speed ratio λ_t									
	1	2	3	4	5	6	7	8	9	10
∞	0.432	0.356	0.282	0.231	0.195	0.168	0.147	0.131	0.118	0.108
1000	0.431	0.355	0.281	0.230	0.194	0.167	0.146	0.130	0.117	0.107
900	0.431	0.355	0.281	0.230	0.194	0.167	0.146	0.130	0.117	0.107
800	0.430	0.354	0.281	0.230	0.193	0.167	0.146	0.130	0.117	0.107
700	0.430	0.354	0.281	0.230	0.193	0.167	0.146	0.130	0.117	0.106
600	0.430	0.354	0.281	0.230	0.193	0.166	0.146	0.130	0.117	0.106
500	0.430	0.354	0.281	0.229	0.193	0.166	0.146	0.129	0.116	0.106
400	0.429	0.353	0.280	0.229	0.192	0.166	0.145	0.129	0.116	0.105
300	0.428	0.352	0.279	0.228	0.192	0.165	0.144	0.128	0.115	0.105
200	0.427	0.351	0.278	0.226	0.190	0.163	0.143	0.127	0.114	0.103
100	0.422	0.346	0.273	0.222	0.186	0.159	0.138	0.122	0.109	0.099
90	0.421	0.345	0.272	0.221	0.185	0.158	0.137	0.121	0.108	0.098
80	0.419	0.344	0.271	0.220	0.183	0.157	0.136	0.120	0.107	0.097
70	0.418	0.342	0.269	0.218	0.182	0.155	0.135	0.119	0.106	0.095
60	0.415	0.340	0.267	0.216	0.180	0.153	0.133	0.116	0.104	0.093
50	0.412	0.336	0.264	0.213	0.177	0.150	0.130	0.114	0.101	0.090
40	0.407	0.332	0.259	0.208	0.172	0.146	0.125	0.109	0.096	0.086
30	0.399	0.324	0.252	0.201	0.165	0.138	0.118	0.102	0.089	0.078
20	0.383	0.308	0.237	0.186	0.150	0.124	0.103	0.087	0.074	0.064
10	0.337	0.263	0.193	0.143	0.107	0.080	0.060	0.044	0.031	0.020

9.2.4 Thrust performance calculation

After consideration of blade tip loss, blade element axial gross thrust is obtained by Formulas (9.7) - (9.9) based on Figure 2.2 is

$$dT = dL_u + dD_u = dL\cos\varphi + dD\sin\varphi = \frac{1}{2}\rho CC_L v^2 \cos\varphi\, dr + \frac{1}{2}\rho CC_D v^2 \sin\varphi\, dr$$

$$= \frac{1}{2}\rho CC_L \cdot U^2 \left[\left(\lambda + \frac{\overline{ag}}{\lambda f}\right)^2 + g^2\right] \frac{\lambda + \dfrac{\overline{ag}}{\lambda f}}{\sqrt{\left(\lambda + \dfrac{\overline{ag}}{\lambda f}\right)^2 + g^2}}\, dr$$

$$+ \frac{1}{2}\rho CC_D \cdot U^2 \left[\left(\lambda + \frac{\overline{ag}}{\lambda f}\right)^2 + g^2\right] \frac{g}{\sqrt{\left(\lambda + \dfrac{\overline{ag}}{\lambda f}\right)^2 + g^2}}\, dr$$

$$= \frac{1}{2}\rho U^2 C \left[\left(\lambda + \frac{\overline{ag}}{\lambda f}\right)C_L + gC_D\right]\sqrt{\left(\lambda + \frac{\overline{ag}}{\lambda f}\right)^2 + g^2}\, dr$$

(9.28)

In respect to wind turbine composed of B blades, substitute relative chord Formula (9.13) into Formula (9.28) for integration to obtain lift coefficient.

$$C_T = \frac{B}{\dfrac{1}{2}\rho U^2 \pi R^2}\int_R \frac{1}{2}\rho U^2 C\left[\left(\lambda + \frac{\overline{ag}}{\lambda f}\right)C_L + gC_D\right]\sqrt{\left(\lambda + \frac{\overline{ag}}{\lambda f}\right)^2 + g^2}\, dr$$

$$= \frac{B}{\pi}\int_R \left(\frac{C}{R}\right)\cdot\left[\left(\lambda + \frac{\overline{ag}}{\lambda f}\right)C_L + gC_D\right]\sqrt{\left(\lambda + \frac{\overline{ag}}{\lambda f}\right)^2 + g^2}\, d\left(\frac{r}{R}\right)$$

$$= \frac{B}{\pi}\int_0^1 \frac{8\pi}{B}\cdot\frac{\overline{a}(1-\overline{a})x}{\left[\left(\lambda_t x + \dfrac{\overline{ag}}{\lambda_t xf}\right)C_L + gC_D\right]\sqrt{\left(\lambda + \dfrac{\overline{ag}}{\lambda f}\right)^2 + g^2}}\cdot\left[\left(\lambda + \frac{\overline{ag}}{\lambda f}\right)C_L + gC_D\right]$$

$$\sqrt{\left(\lambda + \frac{\overline{ag}}{\lambda f}\right)^2 + g^2}\, dx = \int_0^1 8\overline{a}(1-\overline{a})x\, dx$$

(9.29)

Carry out numerical integration, thrust coefficient change trend after correction of blade tip loss is shown as dot matrix in Figure 9.8, and thrust coefficient curve that blade tip loss is not considered has been drawn in the figure.

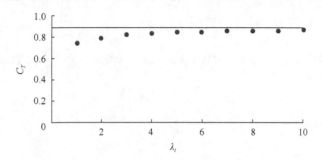

Figure 9.8   Thrust coefficient change trend after correction of blade tip loss

It can be seen in the figure that when tip speed ratio is smaller, limited blade number significantly reduces the thrust coefficient.

9.3   Summary of this chapter

This chapter discussed reasons for blade tip loss, distinguished blade tip local velocity induction factor and mean velocity induction factor and applicability of blade element theory and momentum theory; in a new environment with blade tip loss, blade chord formula is deduced by blade element theory and momentum theory to give another tangent method and example for chord simplification, meanwhile, inflow angle, attack angle and twist change formulas are deduced and twist correction method is given in combination with example. Studies have shown that blade tip loss has small influence on inflow angle at tip part, and great influence on chord; chord must be reduced to 0 at blade tip.

This chapter deduced power, torque, lift and thrust coefficient formulas under the condition with blade tip loss, and carried out numerical integration operation in combination with examples and obtained mass performance data that are related to tip speed ratio and lift drag ratio; these data are closest with actual condition, which is the most valuable goal by designed wind turbine. Numerical integration result for wind turbine power, torque, lift and thrust coefficient showed that blade tip loss significantly reduced various performances of wind turbine, which should not be ignored.

Chapter 10   Practical Blade Structure Design

Practical wind turbine blade structure should be reasonably simplified based on ideal blade structure, with simplification principle is to reduce performance loss to the full extent under the case of meeting mechanical property and in favor of process and manufacturing. This chapter will discuss steps and method of blade structure design under the guidance of the principle, and derive relevant formulas of blade structure element. Based on blade performance designed based on method in this chapter, it will be checked according to the methods described in the next chapter.

10.1   Determination of airfoil and optimum attack angle

Airfoil is selected during design. For power performance of wind turbine, lift drag ratio for airfoil is the important performance, the larger for lift drag ratio, the better for power performance. Selection for airfoil also involves in structure requirement, and various elements should be considered.

Once airfoil is determined, the attack angle at maximum lift drag ratio will be calculated based on lift coefficient and drag coefficient, which will be called as optimum attack angle; Section 3.2 has proved that the optimum attack angle is the attack angle at maximum blade element power. If lift drag ratio changed slightly, the smaller for optimal attack angle of airfoil, the better for it; tip speed ratio can be designed enough large to reduce blade size and weight. The next chapter will discuss the influence of optimal attack angle to design tip speed ratio.

10.2   Determination of design tip speed ratio

Design tip speed ratio is a fixed constant value given during design. Design tip speed ratio is neither too small nor too large. Too small for the ratio will cause excessive width and weight, increasing number of blade and too large of gear speed, resulting in much difficulties during wind turbine production and installation. Design tip speed ratio also can not be too large, if so, blade turning speed is too fast and centrifugal force is large, blade may vibrate and the tip may whistle.

When tip speed ratio is greater than a certain value, inflow angle at the tip ($x=1$) may be less than optimal attack angle of tip airfoil (attack angle with maximum lift drag ratio); since twist can not be negative value (otherwise it will reduce seriously starting performance of wind turbine), and blade fails to work at maximum lift drag ratio, so power will reduce.

To obtain larger tip speed ratio at maximum power, set inflow angle φ at the blade tip equal to or greater than optimal attack angle α_b for tip airfoil (i.e., twist at the blade tip is exactly equal to 0 or greater than 0), that is

$$\tan \varphi \geqslant \tan \alpha_b \qquad (10.1)$$

From Figures 2.2 and 2.3, inflow angle φ can be calculated by the formula

$$\tan \varphi = \frac{u}{w} = \frac{1-a}{1+\dfrac{a(1-a)}{\lambda^2}} \cdot \frac{1}{\lambda} = \frac{6\lambda}{9\lambda^2+2} = \frac{6\lambda_t x}{9\lambda_t^2 x^2 + 2} \qquad (10.2)$$

$x=1$ at the blade tip, so

$$\frac{6\lambda_t}{9\lambda_t^2+2} \geqslant \tan \alpha_b \qquad (10.3)$$

Solve

$$\lambda_{t\,max} \leqslant \frac{1+\sqrt{1-2\tan^2 \alpha_b}}{3\tan \alpha_b} \approx \frac{37.7}{\alpha_b} \qquad (10.4)$$

The unit for optimal attack angle in the formula is degree. Set $x=1$ to obtain tip speed value at the tip with optimal attack angle, relation curve with attack angle is shown in Figure 10.1.

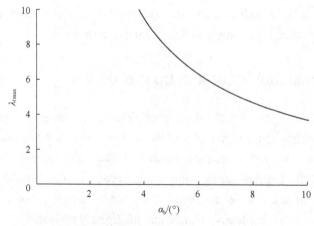

Figure 10.1　Function relationship between maximum design tip speed ratio and optimal attack angle

If optimal attack angle for blade tip airfoil is 6.5°, twist at the tip just decreases to 0 when design tip speed ratio is 5.8, as shown in Figure 10.2.

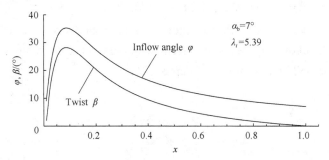

Figure 10.2　Example of influence of optimal attack angle and tip speed ratio on twist

Some airfoils enable to produce a certain lift coefficient when attack angle is 0°, these airfoils generally have smaller optimal attack angle. NACA 4412 airfoil, for example, when Reynolds number is about 5×10^5, lift coefficient for 0° attack angle is about 0.4, optimal attack angle at maximum lift drag ratio is 3°, lift coefficient under this condition is about 0.7; through calculation in Formula (10.4), design tip speed ratio for wind turbine with this airfoil can theoretically reach 12.5.

Common range for design tip speed ratio of modern wind turbine is 5-10, and the optimum attack angle, blade production cost, installation difficulty and blade tip noise should be considered during design.

10.3　Straight chord design

It can be seen from Figure 9.2, chord at the tip after correction decreases to 0, shapes at other position was basically unchanged. This chord curve is also complex, and it is not good for process and making, main body must be simplified. This chapter will still use tangent method, and illustrate in combination with above-mentioned airfoil data.

Set optimal attack angle of the airfoil α_b as 3.5°, lift coefficient of the angle is 0.85, drag coefficient is 0.016. Set the airfoil unchanged along span, and number of blade for wind turbine $B=3$, design tip speed ratio $\lambda_t = 6$, and substitute these parameters into Formula (9.13), chord curve shape of relative chord before and after correction of blade tip loss is shown in Figure 9.2.

Now the chord curve should be simplified by section. Chord curve at the tip zone

$(0.80 \leqslant x \leqslant 1)$ keeps unchanged; tangent line is made at $x=0.8$ in body zone $(2 \leqslant x \leqslant 0.8)$, as shown in Figure 10.3, chord curve is replaced with straight line; blade root cylinder is set at blade root zone $(0 \leqslant x \leqslant 0.1)$, with diameter of 0.03; transition zone is set between root cylinder and main body zone, with the interval of $(0.1 \leqslant x \leqslant 0.2)$, chord curve is replaced with straight line temporarily (smooth transition method will be described in Chapter 12).

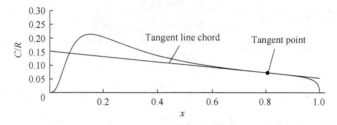

Figure 10.3　Tangent simplification method of chord curve

Chord length at tangent point that $x=0.8$ is 0.072,9, slope is $-0.097,4$, thus tangent equation is

$$\frac{C}{R} = -0.097,4x + 0.150,6 \tag{10.5}$$

When $x=0.2$, relative chord obtained by tangent equation is 0.131,1, and diameter for blade root cylinder is 0.03, therefore straight line equation for transition interval is

$$\frac{C}{R} = 1.011,0x - 0.071,1 \tag{10.6}$$

Blade shape after simplification is shown as broken line in Figure 10.4.

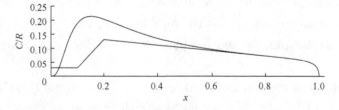

Figure 10.4　Synthetic simplification scheme for chord curve

It can be seen that this simplification method kept blade shape at high power zone of blade exterior, cut large amount of material at low power zone of blade root, and reduced efficiency, but blade size, weight and manufacturing difficulty decreased dramatically, blade root cylinder also improved blade strength.

10.4 Twist correction and treatment

As mentioned above, if attack angle or twist is not corrected after chord is changed, blade will no longer conform to blade element-momentum theorem, and it may not work under design condition, therefore it is necessary to re-adjust distribution of lift coefficient along span, with the method that lift coefficient conforms to chord formula derived by blade element-momentum theorem, but back calculation for lift coefficient is required, and then attack angle and twist are calculated. From Formula (9.13), lift coefficient for ignoring drag coefficient is

$$C_L(x) = \frac{8\pi}{B} \frac{\bar{a}(1-\bar{a})x}{\left(\dfrac{C}{R}\right)\left(\lambda_t x + \dfrac{\bar{a}g}{\lambda_t xf}\right)\sqrt{\left(\lambda_t x + \dfrac{\bar{a}g}{\lambda_t xf}\right)^2 + g^2}} \qquad (10.7)$$

Relative chord C/R can be any expression obtained by using any simplified method.

Lift coefficient distribution formula at tangent zone $(0.2 \leqslant x \leqslant 0.8)$ is

$$C_L(x) = \frac{8\pi}{B} \frac{\bar{a}(1-\bar{a})x}{(-0.097,4x+0.150,6)\left(\lambda_t x + \dfrac{\bar{a}g}{\lambda_t xf}\right)\sqrt{\left(\lambda_t x + \dfrac{\bar{a}g}{\lambda_t xf}\right)^2 + g^2}} \qquad (10.8)$$

Each airfoil meets distribution requirements of above lift coefficient by changing twist or attack angle. For example, if relationship between lift and attack angle for above airfoil is

$$C_L = 0.12(\alpha + 3.6) \qquad (10.9)$$

The unit for attack angle in the formula is degree. The expression for attack angle distributed along span direction at tangent zone is

$$\alpha(x) = \frac{C_L(x)}{0.12} - 3.6$$

$$= \frac{\frac{200\pi}{3B} \cdot \bar{a}(1-\bar{a})x}{(-0.097,4x+0.150,6)\left(\lambda_q x + \frac{\bar{a}g}{\lambda_q xf}\right)\sqrt{\left(\lambda_q x + \frac{\bar{a}g}{\lambda_q xf}\right)^2 + g^2}} - 3.6 \quad (10.10)$$

With inflow angle and attack angle, twist can be obtained immediately. Since the sum of twist and attack angle is equal to inflow angle, twist obtained at tangent zone by Formula (9.14) and above formula is

$$\beta(x) = \varphi(x) - \alpha(x)$$

$$= \frac{180}{\pi} \arctan \frac{gf\lambda_q x}{f\lambda_q^2 x^2 + \bar{a}g} \quad (10.11)$$

$$- \frac{\frac{200\pi}{3B} \cdot \bar{a}(1-\bar{a})x}{(-0.097,4x+0.150,6)\left(\lambda_q x + \frac{\bar{a}g}{\lambda_q xf}\right)\sqrt{\left(\lambda_q x + \frac{\bar{a}g}{\lambda_q xf}\right)^2 + g^2}} + 3.6$$

Chord at blade tip zone $(0.8 \leqslant x \leqslant 1)$ is not changed, and adjustment for twist is not required. Substitute Formulas (9.1) and (9.2) into Formula (10.11), blade twist function graph finally adjusted is shown in Figure 10.5.

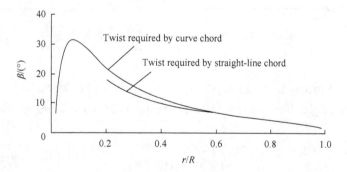

Figure 10.5 Curves for blade inflow angle, attack angle and twist for simplification

Figure 10.5 showed that when chord becomes small, required twist will decrease. Only this equal adjustment is made, blade will not deviate predetermined design condition.

It pointed out in Section 8.1, power coefficient within blade root and transition zone [0, 0.2] only occupies about 1.5%, so it is ignored here and no longer calculated.

10.5 Practical blade shape design

Three elements for blade structure including airfoil, chord and twist were determined in Sections 10.1-10.4.

Airfoil is selected; chord has been simplified: straight line chord is obtained at main body by tangent method, blade root has been set, transition zone has been set between blade root and straight line chord, curve chord is still used at blade tip part (smooth transition among chord part will be discussed in Chapter 12); twist required by straight line chord can be obtained through calculation.

Obviously, airfoil, chord and twist function can be obtained by above steps (although may be segmented function along span), with these three functions, blade function can be structured, and solid figure for practical blade can be obtained by the method of generating blade function graph. Specific method for establishing blade function and obtaining blade stereo image will be described in detail in Chapter 12, here an example for three-dimensional figure generated by blade function is given only (Figure 10.6).

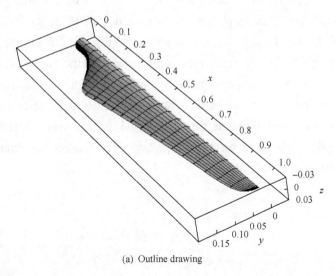

(a) Outline drawing

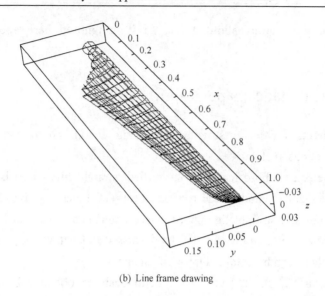

(b) Line frame drawing

Figure 10.6　Example of three-dimensional figure generated by blade function

10.6　Summary of this Chapter

It is difficult to produce ideal blade, and strength fails to meet actual fluid environment, so transformation must be made for actual use. For this purpose, this chapter discussed airfoil selection from the perspective of blade structure design to determine airfoil lift coefficient, drag coefficient and optimum attack angle; gave basic method to determine design tip speed ratio range; proposed the method that generates straight line chord by using tangent line at blade tip part; for the propose of making straight-line chord conform to blade element-momentum theory, re-calculated blade twist to obtain new twist function formula. Blade structure and performance are determined based on above elements, which lays a foundation for practical blade performance calculation in Chapter 11.

Chapter 11 Practical Wind Turbine Performance Calculation

Wind turbine performance can be obtained by integral calculation. If blade chord, twist or lift and drag coefficient are segmented function, they must be calculated by segment during integration. Specific calculation method for performance is discussed now based on Chapter 10.

11.1 Power performance calculation

After consideration of blade tip loss, the blade element power obtained by Formulas (9.7) - (9.9) based on Figure 2.2 is

$$
\begin{aligned}
dP &= \omega r\, dF = \lambda U (dL \sin\varphi - dD \cos\varphi) \\
&= \frac{1}{2}\rho CC_L \lambda U v^2 \sin\varphi\, dr - \frac{1}{2}\rho CC_D v^2 \lambda U \cos\varphi\, dr \\
&= \frac{1}{2}\rho CC_L \lambda U \cdot U^2 \left[\left(\lambda + \frac{\bar{a}g}{\lambda f}\right)^2 + g^2\right] \frac{g}{\sqrt{\left(\lambda + \frac{\bar{a}g}{\lambda f}\right)^2 + g^2}}\, dr \\
&\quad - \frac{1}{2}\rho CC_D \lambda U \cdot U^2 \left[\left(\lambda + \frac{\bar{a}g}{\lambda f}\right)^2 + g^2\right] \frac{\lambda + \frac{\bar{a}g}{\lambda f}}{\sqrt{\left(\lambda + \frac{\bar{a}g}{\lambda f}\right)^2 + g^2}}\, dr \\
&= \frac{1}{2}\rho U^3 C\lambda \left[gC_L - \left(\lambda + \frac{\bar{a}g}{\lambda f}\right)C_D\right]\sqrt{\left(\lambda + \frac{\bar{a}g}{\lambda f}\right)^2 + g^2}\, dr
\end{aligned} \quad (11.1)
$$

In respect to wind turbine composed of B blades, substitute relative chord Formula (9.13) into Formula (11.1) for integration (influence of drag coefficient on chord has been ignored) to obtain power coefficient.

$$C_P = \frac{B}{\frac{1}{2}\rho U^3 \pi R^2} \int_R \frac{1}{2} \rho U^3 C \lambda \left[gC_L - \left(\lambda + \frac{\bar{a}g}{\lambda f}\right) C_D \right] \sqrt{\left(\lambda + \frac{\bar{a}g}{\lambda f}\right)^2 + g^2} \, dr$$

$$= \frac{B}{\pi} \int_R \left(\frac{C}{R}\right) \cdot \lambda C_L \left[g - \left(\lambda + \frac{\bar{a}g}{\lambda f}\right) \frac{C_D}{C_L} \right] \sqrt{\left(\lambda + \frac{\bar{a}g}{\lambda f}\right)^2 + g^2} \, d\left(\frac{r}{R}\right)$$

$$= \frac{B}{\pi} \int_R \frac{8\pi}{B} \frac{\bar{a}(1-\bar{a})x}{C_L \left(\lambda_t x + \frac{\bar{a}g}{\lambda_t x f}\right) \sqrt{\left(\lambda_t x + \frac{\bar{a}g}{\lambda_t x f}\right)^2 + g^2}} \quad (11.2)$$

$$\lambda C_L \left[g - \left(\lambda + \frac{\bar{a}g}{\lambda f}\right) \frac{C_D}{C_L} \right] \sqrt{\left(\lambda + \frac{\bar{a}g}{\lambda f}\right)^2 + g^2} \, d\left(\frac{r}{R}\right)$$

$$= \int_R 8\bar{a}(1-\bar{a})\lambda_t x^2 \left(\frac{g}{\lambda_t x + \frac{\bar{a}g}{\lambda_t x f}} - \frac{C_D}{C_L} \right) dx$$

This is a general formula to calculate power coefficient. When drag coefficient is 0, or tip speed ratio is infinitely great, the formula will be transformed to maximum power performance formula of practical wind turbine, the same as Formula (9.18). It is observed from Formula (11.2), integrand for actual wind turbine power performance formula is difference between drag-lift ratio and integrand for actual wind turbine maximum power performance formula, obviously, the higher for lift drag ratio, the higher for power coefficient of actual wind turbine.

To further describe calculation method of performance, examples for data in Chapter 10 are taken, main calculation conditions: relationship among airfoil lift, drag coefficient and attack angle are

$$C_L = 0.12(\alpha + 3.6) \quad (11.3)$$

$$C_D = 0.012 + 1.052(\pi\alpha/180)^2 \quad (11.4)$$

The unit for attack angle in the formula is degree. Set $\partial(C_L/C_D)/\partial\alpha = 0$ to obtain optimum attack angle of maximum lift drag ratio α_b as 3.5°, lift coefficient of the angle is 0.85, drag coefficient is 0.016. Other calculation condition: set the airfoil unchanged along span, and number of blade for wind turbine $B=3$, design tip speed ratio $\lambda_t = 6$.

Since chord and twist are segmented functions, integral calculation is required by segment. Substitute Formulas (9.1), (9.2), (9.6), (11.3), (11.4) and attack angle

Formula (10.10) into Formula (11.2), and power coefficient at tangent zone ($0.2 \leqslant x \leqslant 0.8$) by numerical integration is

$$C_{P1} = \int_{0.2}^{0.8} 8\bar{a}(1-\bar{a})\lambda_t x^2 \left(\frac{g}{\lambda_t x + \frac{\bar{a}g}{\lambda_t xf}} - \frac{C_D}{C_L} \right) dx$$

$$= \int_{0.2}^{0.8} 8\bar{a}(1-\bar{a})\lambda_t x^2 \left(\frac{g}{\lambda_t x + \frac{\bar{a}g}{\lambda_t xf}} - \frac{0.012 + 1.052(\pi\alpha/180)^2}{0.12(\alpha+3.6)} \right) dx$$

$$= \int_{0.2}^{0.8} 48\bar{a}(1-\bar{a})x^2 \left(\frac{g}{6x + \frac{\bar{a}g}{6xf}} - \frac{0.012 + 1.052\left\{\pi\left[\dfrac{200\pi}{9} - \dfrac{\bar{a}(1-\bar{a})x}{(-0.097,4x+0.150,6)\left(6x+\dfrac{\bar{a}g}{6xf}\right)\sqrt{\left(6x+\dfrac{\bar{a}g}{6xf}\right)^2 + g^2}} - 3.6\right]/180\right\}^2}{0.12 \cdot \dfrac{200\pi}{9} - \dfrac{\bar{a}(1-\bar{a})x}{(-0.097,4x+0.150,6)\left(6x+\dfrac{\bar{a}g}{\lambda_t xf}\right)\sqrt{\left(6x+\dfrac{\bar{a}g}{6xf}\right)^2 + g^2}}} \right) dx$$

$$= 0.310,7$$

(11.5)

Attack angle at blade tip zone is optimal attack angle (3.5°), and power coefficient is

$$C_{P2} = \int_{0.8}^{1} 8\bar{a}(1-\bar{a})\lambda_t x^2 \left(\frac{g}{\lambda_t x + \frac{\bar{a}g}{\lambda_t xf}} - \frac{C_D}{C_L} \right) dx$$

$$= \int_{0.8}^{1} 8\bar{a}(1-\bar{a})\lambda_t x^2 \left(\frac{g}{\lambda_t x + \frac{\bar{a}g}{\lambda_t xf}} - \frac{0.012 + 1.052(3.5\pi/180)^2}{0.12(3.5+3.6)} \right) dx \quad (11.6)$$

$$= 0.147,5$$

Power coefficient for overall blade is

$$C_P = C_{P1} + C_{P2} = 0.310,7 + 0.147,5 = 0.458,2 \quad (11.7)$$

11.2 Torque performance calculation

After consideration of blade tip loss, the blade element torque obtained by Formulas (9.7)-(9.9) based on Figure 2.2 is

$$dM = r\,dF = r(dL\sin\varphi - dD\cos\varphi)$$

$$= \frac{1}{2}\rho CC_L v^2 \sin\varphi \cdot r\,dr - \frac{1}{2}\rho CC_D v^2 \cos\varphi \cdot r\,dr$$

$$= \frac{1}{2}\rho CC_L \cdot U^2 \left[\left(\lambda + \frac{\bar{a}g}{\lambda f}\right)^2 + g^2\right] \frac{g}{\sqrt{\left(\lambda + \frac{\bar{a}g}{\lambda f}\right)^2 + g^2}} r\,dr$$

$$- \frac{1}{2}\rho CC_D \cdot U^2 \left[\left(\lambda + \frac{\bar{a}g}{\lambda f}\right)^2 + g^2\right] \frac{\lambda + \frac{\bar{a}g}{\lambda f}}{\sqrt{\left(\lambda + \frac{\bar{a}g}{\lambda f}\right)^2 + g^2}} r\,dr \qquad (11.8)$$

$$= \frac{1}{2}\rho U^2 C \left[gC_L - \left(\lambda + \frac{\bar{a}g}{\lambda f}\right)C_D\right]\sqrt{\left(\lambda + \frac{\bar{a}g}{\lambda f}\right)^2 + g^2} \cdot r\,dr$$

In respect to wind turbine composed of B blades, substitute relative chord Formula (9.13) into Formula (11.8) for integration (influence of drag coefficient on chord has been ignored) to obtain torque coefficient.

$$C_M = \frac{B}{\frac{1}{2}\rho U^2 \pi R^3} \int_R \frac{1}{2}\rho U^2 C\left[gC_L - \left(\lambda + \frac{\bar{a}g}{\lambda f}\right)C_D\right]\sqrt{\left(\lambda + \frac{\bar{a}g}{\lambda f}\right)^2 + g^2} \cdot r\,dr$$

$$= \frac{B}{\pi}\int_R \left(\frac{C}{R}\right)\left[gC_L - \left(\lambda + \frac{\bar{a}g}{\lambda f}\right)C_D\right]\sqrt{\left(\lambda + \frac{\bar{a}g}{\lambda f}\right)^2 + g^2} \cdot \left(\frac{r}{R}\right)d\left(\frac{r}{R}\right)$$

$$= \frac{B}{\pi}\int_0^1 \frac{8\pi}{B} \frac{\bar{a}(1-\bar{a})x}{C_L\left(\lambda_t x + \frac{\bar{a}g}{\lambda_t xf}\right)\sqrt{\left(\lambda + \frac{\bar{a}g}{\lambda f}\right)^2 + g^2}}\left[gC_L - \left(\lambda_t x + \frac{\bar{a}g}{\lambda_t xf}\right)C_D\right]\sqrt{\left(\lambda + \frac{\bar{a}g}{\lambda f}\right)^2 + g^2} \cdot x\,dx$$

$$= \int_0^1 8\bar{a}(1-\bar{a})x^2 \left(\frac{g}{\lambda_t x + \frac{\bar{a}g}{\lambda_t xf}} - \frac{C_D}{C_L}\right)dx$$

$$\qquad (11.9)$$

This is a general formula to calculate torque coefficient. When drag coefficient is 0, or tip speed ratio is infinitely great, the formula will be transformed to maximum torque performance formula of practical wind turbine, the same as Formula (9.22).

Since chord and twist are segmented functions, integral calculation is required by segment. Substitute Formulas (9.1), (9.2), (9.6), (11.3), (11.4) and attack angle Formula (10.10) into Formula (11.9), and torque coefficient at tangent zone $(0.2 \leqslant x \leqslant 0.8)$ by numerical integration is

$$C_{M1} = \int_{0.2}^{0.8} 8\bar{a}(1-\bar{a})x^2 \left(\frac{g}{\lambda_t x + \dfrac{\overline{ag}}{\lambda_t xf}} - \frac{C_D}{C_L} \right) dx$$

$$= \int_{0.2}^{0.8} 8\bar{a}(1-\bar{a})x^2 \left(\frac{g}{\lambda_t x + \dfrac{\overline{ag}}{\lambda_t xf}} - \frac{0.012 + 1.052(\pi\alpha/180)^2}{0.12(\alpha+3.6)} \right) dx$$

$$= \int_{0.2}^{0.8} 48\bar{a}(1-\bar{a})x^2 \left\{ \frac{g}{6x + \dfrac{\overline{ag}}{6xf}} - \frac{0.012 + 1.052 \left\{ \pi \left[\dfrac{200\pi}{9} \cdot \dfrac{\bar{a}(1-\bar{a})x}{(-0.097,4x+0.150,6)\left(6x+\dfrac{\overline{ag}}{6xf}\right)\sqrt{\left(6x+\dfrac{\overline{ag}}{6xf}\right)^2 + g^2}} - 3.6 \right] / 180 \right\}^2}{0.12 \cdot \dfrac{200\pi}{9} \cdot \dfrac{\bar{a}(1-\bar{a})x}{(-0.097,4x+0.150,6)\left(6x+\dfrac{\overline{ag}}{\lambda_t xf}\right)\sqrt{\left(6x+\dfrac{\overline{ag}}{6xf}\right)^2 + g^2}}} \right\} dx$$

$$= 0.051,8 \tag{11.10}$$

Attack angle at blade tip zone is optimal attack angle (3.5°), and the torque coefficient is

$$C_{M2} = \int_{0.8}^{1} 8\bar{a}(1-\bar{a})x^2 \left(\frac{g}{\lambda_t x + \dfrac{\overline{ag}}{\lambda_t xf}} - \frac{C_D}{C_L} \right) dx$$

$$= \int_{0.8}^{1} 8\bar{a}(1-\bar{a})x^2 \left(\frac{g}{\lambda_t x + \dfrac{\overline{ag}}{\lambda_t xf}} - \frac{0.012 + 1.052(3.5\pi/180)^2}{0.12(3.5+3.6)} \right) dx \tag{11.11}$$

$$= 0.024,6$$

Torque coefficient for overall blade is

$$C_M = C_{M1} + C_{M2}$$
$$= 0.051,8 + 0.024,6 = 0.076,4 \tag{11.12}$$

11.3 Lift performance calculation

After consideration of blade tip loss, blade element lift obtained by Formulas (9.7)-(9.9) based on Figure 2.2 is

$$dF = dL\sin\varphi - dD\cos\varphi$$

$$= \frac{1}{2}\rho C C_L v^2 \sin\varphi dr - \frac{1}{2}\rho C C_D v^2 \cos\varphi dr$$

$$= \frac{1}{2}\rho C C_L \cdot U^2 \left[\left(\lambda + \frac{\bar{a}g}{\lambda f}\right)^2 + g^2\right] \frac{g}{\sqrt{\left(\lambda + \frac{\bar{a}g}{\lambda f}\right)^2 + g^2}} dr$$

$$-\frac{1}{2}\rho C C_D \cdot U^2 \left[\left(\lambda + \frac{\bar{a}g}{\lambda f}\right)^2 + g^2\right] \frac{\lambda + \frac{\bar{a}g}{\lambda f}}{\sqrt{\left(\lambda + \frac{\bar{a}g}{\lambda f}\right)^2 + g^2}} dr \quad (11.13)$$

$$= \frac{1}{2}\rho U^2 C \left[gC_L - \left(\lambda + \frac{\bar{a}g}{\lambda f}\right)C_D\right] \sqrt{\left(\lambda + \frac{\bar{a}g}{\lambda f}\right)^2 + g^2} \, dr$$

$$C_F = \frac{B}{\frac{1}{2}\rho U^2 \pi R^2} \int_R \frac{1}{2}\rho U^2 C \left[gC_L - \left(\lambda + \frac{\bar{a}g}{\lambda f}\right)C_D\right] \sqrt{\left(\lambda + \frac{\bar{a}g}{\lambda f}\right)^2 + g^2} \, dr$$

$$= \frac{B}{\pi} \int_R \left(\frac{C}{R}\right)\left[gC_L - \left(\lambda + \frac{\bar{a}g}{\lambda f}\right)C_D\right]\sqrt{\left(\lambda + \frac{\bar{a}g}{\lambda f}\right)^2 + g^2} \, d\left(\frac{r}{R}\right)$$

$$= \frac{B}{\pi}\int_0^1 \frac{8\pi}{B} \frac{\bar{a}(1-\bar{a})x}{C_L\left(\lambda_1 x + \frac{\bar{a}g}{\lambda_1 x f}\right)\sqrt{\left(\lambda + \frac{\bar{a}g}{\lambda f}\right)^2 + g^2}} \left[gC_L - \left(\lambda_1 x + \frac{\bar{a}g}{\lambda_1 x f}\right)C_D\right]\sqrt{\left(\lambda + \frac{\bar{a}g}{\lambda f}\right)^2 + g^2} \, dx$$

$$= \int_0^1 8\bar{a}(1-\bar{a})x\left[\frac{g}{\lambda_1 x + \frac{\bar{a}g}{\lambda_1 x f}} - \frac{C_D}{C_L}\right]dx$$

(11.14)

In respect to wind turbine composed of B blades, substitute relative chord Formula (9.13) into Formula (11.14) for integration (influence of drag coefficient on chord has been ignored) to obtain lift coefficient.

Chapter 11 Practical Wind Turbine Performance Calculation

This is a general formula to calculate lift coefficient. When drag coefficient is 0, or tip speed ratio is infinitely great, the formula will be transformed to maximum lift performance formula of practical wind turbine, the same as Formula (9.26).

Since chord and twist are segmented functions, integral calculation is required by segment. Substitute Formulas (9.1), (9.2), (9.6), (11.3), (11.4) and attack angle Formula (10.10) into Formula (11.9), and lift coefficient at tangent zone ($0.2 \leqslant x \leqslant 0.8$) by numerical integration is

$$C_{F1} = \int_{0.2}^{0.8} 8\bar{a}(1-\bar{a})x \left[\frac{g}{\lambda_i x + \frac{\bar{a}g}{\lambda_i xf}} - \frac{C_D}{C_L} \right] dx$$

$$= \int_{0.2}^{0.8} 8\bar{a}(1-\bar{a})x \left[\frac{g}{\lambda_i x + \frac{\bar{a}g}{\lambda_i xf}} - \frac{0.012 + 1.052(\pi\alpha/180)^2}{0.12(\alpha+3.6)} \right] dx$$

$$= \int_{0.2}^{0.8} 48\bar{a}(1-\bar{a})x \left[\frac{g}{6x + \frac{\bar{a}g}{6xf}} - \frac{0.012 + 1.052\left\{ \pi \left[\frac{200\pi}{9} \frac{\bar{a}(1-\bar{a})x}{(-0.097,4x+0.150,6)\left(6x+\frac{\bar{a}g}{6xf}\right)\sqrt{\left(6x+\frac{\bar{a}g}{6xf}\right)^2+g^2}} - 3.6 \right]/180 \right\}}{0.12 \cdot \frac{200\pi}{9} \frac{\bar{a}(1-\bar{a})x}{(-0.097,4x+0.150,6)\left(6x+\frac{\bar{a}g}{\lambda_i xf}\right)\sqrt{\left(6x+\frac{\bar{a}g}{6xf}\right)^2+g^2}}} \right] dx$$

$$= 0.103,4$$

(11.15)

Attack angle at blade tip zone is optimal attack angle (3.5°), and lift coefficient is

$$C_{F2} = \int_{0.8}^{1} 8\bar{a}(1-\bar{a})x \left[\frac{g}{\lambda_i x + \frac{\bar{a}g}{\lambda_i xf}} - \frac{C_D}{C_L} \right] dx$$

$$= \int_{0.8}^{1} 8\bar{a}(1-\bar{a})x \left[\frac{g}{\lambda_i x + \frac{\bar{a}g}{\lambda_i xf}} - \frac{0.012 + 1.052(3.5\pi/180)^2}{0.12(3.5+3.6)} \right] dx \quad (11.16)$$

$$= 0.027,6$$

Lift coefficient for overall blade is

$$C_F = C_{F1} + C_{F2} = 0.103,7 + 0.027,6 = 0.131,3 \quad (11.17)$$

11.4 Thrust performance calculation

After consideration of blade tip loss, blade element axial gross thrust obtained by Formulas (9.7)-(9.9) based on Figure 2.2 is

$$dT = dL_u + dD_u = dL\cos\varphi + dD\sin\varphi$$

$$= \frac{1}{2}\rho CC_L v^2 \cos\varphi dr + \frac{1}{2}\rho CC_D v^2 \sin\varphi dr$$

$$= \frac{1}{2}\rho CC_L \cdot U^2 \left[\left(\lambda + \frac{\bar{a}g}{\lambda f}\right)^2 + g^2\right] \frac{\lambda + \frac{\bar{a}g}{\lambda f}}{\sqrt{\left(\lambda + \frac{\bar{a}g}{\lambda f}\right)^2 + g^2}} dr$$

$$+ \frac{1}{2}\rho CC_D \cdot U^2 \left[\left(\lambda + \frac{\bar{a}g}{\lambda f}\right)^2 + g^2\right] \frac{g}{\sqrt{\left(\lambda + \frac{\bar{a}g}{\lambda f}\right)^2 + g^2}} dr$$

$$= \frac{1}{2}\rho U^2 C \left[\left(\lambda + \frac{\bar{a}g}{\lambda f}\right) C_L + gC_D\right] \sqrt{\left(\lambda + \frac{\bar{a}g}{\lambda f}\right)^2 + g^2} \, dr \qquad (11.18)$$

In respect to wind turbine composed of B blades, substitute relative chord Formula (9.13) into Formula (11.18) for integration (influence of drag coefficient on chord has been ignored) to obtain thrust coefficient.

$$C_T = \frac{B}{\frac{1}{2}\rho U^2 \pi R^2} \int_R \frac{1}{2}\rho U^2 C \left[\left(\lambda + \frac{\bar{a}g}{\lambda f}\right) C_L + gC_D\right] \sqrt{\left(\lambda + \frac{\bar{a}g}{\lambda f}\right)^2 + g^2} \, dr$$

$$= \frac{B}{\pi} \int_R \left(\frac{C}{R}\right) \left[\left(\lambda + \frac{\bar{a}g}{\lambda f}\right) C_L + gC_D\right] \sqrt{\left(\lambda + \frac{\bar{a}g}{\lambda f}\right)^2 + g^2} \, d\left(\frac{r}{R}\right)$$

$$= \frac{B}{\pi} \int_0^1 \frac{8\pi}{B} \frac{\bar{a}(1-\bar{a})x}{\left[\left(\lambda_q x + \frac{\bar{a}g}{\lambda_q xf}\right) C_L + gC_D\right] \sqrt{\left(\lambda + \frac{\bar{a}g}{\lambda f}\right)^2 + g^2}} \left[\left(\lambda + \frac{\bar{a}g}{\lambda f}\right) C_L + gC_D\right] \sqrt{\left(\lambda + \frac{\bar{a}g}{\lambda f}\right)^2 + g^2} \, dx$$

$$= \int_0^1 8\bar{a}(1-\bar{a})x \, dx$$

$$= 0.853,9$$

$$(11.19)$$

11.5　Summary of this Chapter

　　This chapter derived power, torque, lift and drag performance calculation formula of actual wind turbine composed by the blade by using blade element-momentum theory based on main elements described in Chapter 10, and calculated numerical integration for various performance in combination with examples, and obtained specific performance for power, torque, lift and thrust of blades. Studies have shown that performance of actual blades can be obtained by analytic calculation method; studies also have shown that simplification way for ideal blade (tangent position, straight-line chord length) directly affected blade performance.

Chapter 12　Blade Function Design Methods

If airfoil, chord and twist function have been obtained, blade three-dimensional image can be obtained inevitably by function graph generation way to realize functional design of blade shape. Blade shape is the curved surface with very complex structure, so it is quite difficult to build curved surface function and generate complex curved surface, and there are little relevant research literature. However, building mathematical model for blade has important significance, first, mathematical model enables to reflect essential law of blade structure, people can further deepen understanding to blade function by mathematical model; second, mathematical model is used to carry out functional design of blade: mathematical expression structure or parameters are adjusted to generate various shapes of blade image, so that large quantity work in blade design is completed automatically by mathematical software.

The method by using blade function to generate blade stereo image is known as blade functional design method in this book. This chapter will build general expression of blade function of wind turbine (also called mathematical model of blade) by using three sub-functions including airfoil, chord and twist, and study the method that generates three-dimensional stereogram by mathematical software.

There are two goals to draw blade dimensional stereogram of wind turbine, one is to analyze and model for structural dynamics and fluid dynamics[70], the other one is to process and manufacture. At present, there are many reports[16,71] for blade three-dimensional model design method and examples based on various mapping software, most of them focused on studying single modeling, with common characteristic: discrete coordinate point of airfoil is input into large mapping software, such as Pro/E, SolidWorks, UG by coordinate transformation, and generate blade stereogram by manual fairing or change. Different point of drawing three-dimensional figure by functional design method is to adopt general mathematical software, what is input is function expression, not discrete coordinate, therefore the result obtained is functional image of blade curved surface. Solid figure can be drawn by mathematical software when airfoil profile expression, chord change expression along span and twist change expression along span are given, this is more convenient, rapid and precise than large mapping software, especially making for design analysis and modification.

12.1 Mathematical model for blade of wind turbine

Mathematical model for blade of horizontal axis wind turbine is mathematical expression of reflecting physical blade shape, it can be an independent equation or a group of parameter equation. Blade mathematical model in this book also includes meaning of blade three-dimensional model (stereo image) by using mathematical expression.

12.1.1 Coordinate transformation for airfoil function

Airfoil profile can be expressed as non-dimensional equation of chord C, and general type is

$$z_C = f(y_C) \tag{12.1}$$

z_C and y_C represent to y-coordinate and x-coordinate that is relative to chord C, airfoil coordinate system and airfoil function graph are shown in Figure 12.1.

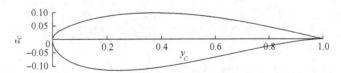

Figure 12.1 Airfoil and airfoil coordinate system example

In studying the whole blade, $Oxyz$ coordinate system (also called blade coordinate system, and see Figure 12.2) is required, y axis and z axis coincide with y_C axis and z_C axis in airfoil coordinate system that twist is 0, y axis is airfoil chord line direction (taking the direction from leading edge to trailing edge as forward), z axis is the same as remote wind speed direction, x axis is vertical to Oyz plane (taking the direction from blade root to tip as forward).

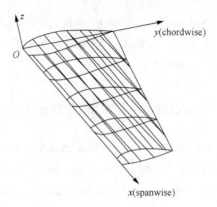

Figure 12.2   Blade coordinate system example

Blade chord C changes along span, and chord function can be expressed as $C=C(x)$. In blade coordinate system, the coordinate of airfoil that is relative to chord C is required to be converted into the coordinate of blade length R. The expression of airfoil x-coordinate transformation is

$$y_R = \frac{C(x)}{R} y_C \tag{12.2}$$

y-coordinate of airfoil surface is required to be converted into the coordinate that is relative to blade length R, with the expression

$$z_R = \frac{C(x)}{R} z_C = \frac{C(x)}{R} f(y_C) \tag{12.3}$$

Thus, blade surface is the curved surface about x, y_C.

12.1.2   Twist rotation transformation

Now blade twist rotation transformation is studied, i.e., making airfoil rotates in three-dimensional space according to twist value. Twist β changes along span, and twist function can be expressed as $\beta = \beta(x)$. From view of x axis (span direction), Ozy plane is unchanged while twist rotates airfoil curve, and rotation transformation formula is[72]

$$\begin{cases} y = y_R \cos\beta - z_R \sin\beta \\ z = y_R \sin\beta + z_R \cos\beta \end{cases} \tag{12.4}$$

Obviously when twist $\beta=0$, $y=y_R$, $z=z_R$. Substitute Formulas (12.2) and (12.3) into Formula (12.4), and the expression of airfoil with twist in three-dimensional space.

$$\begin{cases} y = \dfrac{C(x)}{R} y_C \cos\beta(x) - \dfrac{C(x)}{R} f(y_C)\sin\beta(x) \\ z = \dfrac{C(x)}{R} y_C \sin\beta(x) + \dfrac{C(x)}{R} f(y_C)\cos\beta(x) \end{cases} \quad (12.5)$$

12.1.3 Build blade mathematical model

Eliminate $f(y_C)$ in Formula (12.5) to obtain

$$y_C = \frac{R}{C}(y\cos\beta + z\sin\beta) \quad (12.6)$$

Substitute the formula into the second equation in Formula (12.5) to obtain general equation of blade surface.

$$z = (y\cos\beta + z\sin\beta)\sin\beta + \frac{C}{R}f\left(\frac{R}{C}y\cos\beta + \frac{R}{C}z\sin\beta\right)\cos\beta \quad (12.7)$$

$f()$ represents the function. Independent variable for chord C and twist β in Formula (12.7) is span coordinate x.

It is difficult to express curved surface equation to explicit function way, which is not good for drawing image. To facilitate to draw curved surface image, blade surface equation is considered to be transferred to explicit function way of parameter formula. Formula (12.5) is established for each x_R position (the distance that is relative to R from a point of span to Oyz plane), x_R and y_C can be regarded as parameters, therefore parametric equation for blade surface based on Formula (12.5) is

$$\begin{cases} x = x_R \\ y = \dfrac{C(x_R)}{R} y_C \cos\beta(x_R) - \dfrac{C(x_R)}{R} f(y_C)\sin\beta(x_R) \\ z = \dfrac{C(x_R)}{R} y_C \sin\beta(x_R) + \dfrac{C(x_R)}{R} f(y_C)\cos\beta(x_R) \end{cases} \quad (12.8)$$

Parameter x_R and y_C have specific geometrical significance: x_R refers to relative position of blade in span direction, y_C refers to airfoil chordwise relative position, both change interval for parameters are [0,1]. Blade surface space coordinate (x, y, z)

obtained by the equation set is relative value of blade length R, if multiplying R, actual space coordinate will be obtained.

General Formula (12.7) or parametric Formula (12.8) for blade surface is universal blade mathematical model. It is easy to understand the geometrical significance of Formula (12.8): each spanwise position (the first equation) for blade, the last two equations in equation set represent airfoil shape of cross section at the position, and rotating function and dimensional consistency of the twist.

12.1.4 Example for blade image generation

Generating blade three-dimensional image by functional design method requires substitute specific expression for airfoil function $f(y_C)$, twist β and chord C into blade surface parametric Formula (12.8). Now an ideal blade example is given.

Set airfoil profile expression as

$$f_u(y_C) = 0.2 y_C (1-y_C) + 0.17 y_C^{0.5} (1-y_C)^{1.5} \tag{12.9}$$

$$f_l(y_C) = 0.02 y_C (1-y_C) - 0.37 y_C^{0.5} (1-y_C)^{1.5} \tag{12.10}$$

The shape is shown in Figure 12.1.

Set optimal attack angle of airfoil α as 3.5°, lift coefficient of the attack angle as 0.85. Set airfoil along span remain unchanged, design tip speed ratio of wind turbine $\lambda_t = 6$. Substitute these values into Formula (3.13) to twist expression along span.

$$\beta(x) = \arctan \frac{36x}{324x^2 + 2} - 3.5° \tag{12.11}$$

Substitute above value into Formula (8.8), and ignore the influence of drag coefficient, chord change expression along span is obtained as

$$C(x) = \frac{16\pi R}{27} \frac{x}{0.85\left(6x + \dfrac{2}{54x}\right)\sqrt{\left(6x + \dfrac{2}{54x}\right)^2 + \left(\dfrac{2}{3}\right)^2}} \tag{12.12}$$

Substitute twist Formula (12.11), chord Formula (12.12), upward profile Formula (12.9) into blade function Formula (12.8) to obtain blade curved surface function formed by upward profile:

$$\begin{cases} x = x_R \\ y = \dfrac{16\pi R}{27} \dfrac{x_R y_C}{0.85\left(6x_R + \dfrac{2}{54x_R}\right)\sqrt{\left(6x_R + \dfrac{2}{54x_R}\right)^2 + \left(\dfrac{2}{3}\right)^2}} \cos\left(\arctan\dfrac{36x}{324x^2+2} - 3.5°\right) - \\ \qquad \dfrac{16\pi R}{27} \dfrac{x_R\left[0.2y_C(1-y_C) + 0.17y_C^{0.5}(1-y_C)^{1.5}\right]}{0.85\left(6x_R + \dfrac{2}{54x_R}\right)\sqrt{\left(6x_R + \dfrac{2}{54x_R}\right)^2 + \left(\dfrac{2}{3}\right)^2}} \sin\left(\arctan\dfrac{36x}{324x^2+2} - 3.5°\right) \\ z = \dfrac{16\pi R}{27} \dfrac{x_R y_C}{0.85\left(6x_R + \dfrac{2}{54x_R}\right)\sqrt{\left(6x_R + \dfrac{2}{54x_R}\right)^2 + \left(\dfrac{2}{3}\right)^2}} \sin\left(\arctan\dfrac{36x}{324x^2+2} - 3.5°\right) + \\ \qquad \dfrac{16\pi R}{27} \dfrac{x_R\left[0.2y_C(1-y_C) + 0.17y_C^{0.5}(1-y_C)^{1.5}\right]}{0.85\left(6x_R + \dfrac{2}{54x_R}\right)\sqrt{\left(6x_R + \dfrac{2}{54x_R}\right)^2 + \left(\dfrac{2}{3}\right)^2}} \cos\left(\arctan\dfrac{36x}{324x^2+2} - 3.5°\right) \end{cases}$$
(12.13)

Substitute twist Formula (12.11), chord Formula (12.12) and downward profile Formula (12.10) into blade function Formula (12.8) to obtain blade curved surface function formed by downward profile:

$$\begin{cases} x = x_R \\ y = \dfrac{16\pi R}{27} \dfrac{x_R y_C}{0.85\left(6x_R + \dfrac{2}{54x_R}\right)\sqrt{\left(6x_R + \dfrac{2}{54x_R}\right)^2 + \left(\dfrac{2}{3}\right)^2}} \cos\left(\arctan\dfrac{36x}{324x^2+2} - 3.5°\right) - \\ \qquad \dfrac{16\pi R}{27} \dfrac{x_R\left[0.02y_C(1-y_C) - 0.37y_C^{0.5}(1-y_C)^{1.5}\right]}{0.85\left(6x_R + \dfrac{2}{54x_R}\right)\sqrt{\left(6x_R + \dfrac{2}{54x_R}\right)^2 + \left(\dfrac{2}{3}\right)^2}} \sin\left(\arctan\dfrac{36x}{324x^2+2} - 3.5°\right) \\ z = \dfrac{16\pi R}{27} \dfrac{x_R y_C}{0.85\left(6x_R + \dfrac{2}{54x_R}\right)\sqrt{\left(6x_R + \dfrac{2}{54x_R}\right)^2 + \left(\dfrac{2}{3}\right)^2}} \sin\left(\arctan\dfrac{36x}{324x^2+2} - 3.5°\right) + \\ \qquad \dfrac{16\pi R}{27} \dfrac{x_R\left[0.02y_C(1-y_C) - 0.37y_C^{0.5}(1-y_C)^{1.5}\right]}{0.85\left(6x_R + \dfrac{2}{54x_R}\right)\sqrt{\left(6x_R + \dfrac{2}{54x_R}\right)^2 + \left(\dfrac{2}{3}\right)^2}} \cos\left(\arctan\dfrac{36x}{324x^2+2} - 3.5°\right) \end{cases}$$
(12.14)

x_R refers to relative position of blade in span direction, y_C refers to airfoil chordwise relative position, both change interval for parameter is [0, 1]. By means of the blade function, Mathematica software can be used to draw blade stereo image (see Section 12.3 for specific drawing method), and shape is shown in Figure 12.3.

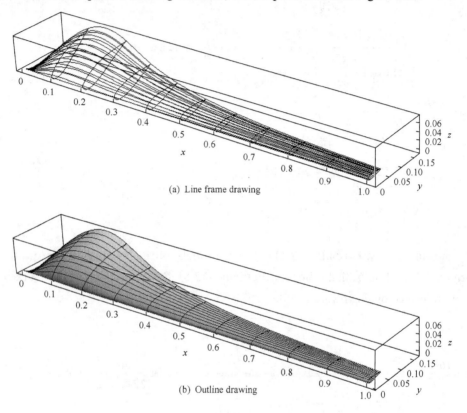

Figure 12.3 Stereo image for ideal blade

If blade root is set midpoint of chord (midpoint of chord is located at x axis), stereo image is shown in Figure 3.4.

Airfoil for ideal blade is selected, and chord and twist are derived theoretically. In practical application, blade root and tip loss correction should be considered for chord and twist, and simplification for processing and manufacturing should be carried out, therefore, ideal blade differs greatly from actual blade shape.

12.2 Smooth transition among airfoils

Important characteristics of ideal blade are that airfoil keeps consistent along span, and twist and chord curve are not segmented along span. Simplified blade and practical blade are different, chord curve is divided into blade tip section, blade root section, chord straight-line section and transition section, in addition, blade is irregular along span, and some blades adopt several airfoils, and smooth transition should be made between adjacent airfoil to allow the surface to smooth as far as possible. Even if a kind of airfoil is used, it will involve in smooth transition between blade root cylinder and blade airfoil.

Transition between blade root cylinder and blade airfoil is realized by spline interpolation[73,74]. Large mapping software uses its own "visual" algorithm to make smoothness check and finishing for blade curved surface, for example in UG appearance molding mode, "curvature comb" and "light and shadow analysis" are used to make analysis and check, and cutting and supplement are made manually to realize smooth transition of curved surface.

This section discussed smooth transition at different space positions by function method.

12.2.1 Sine curve fairing method

Set airfoil transition zone distribute between sections of relative position of span x_1 and x_2, smooth transition is required for all parameters of airfoil. There are many smooth transition methods, this section will discuss the method of connecting parameters at different position by lowest point and highest point by about half cycle sine curve, which is abbreviated as sine curve fairing method.

Set parameters at x_1 and x_2 as P_1 and P_2, and connect these two parameters smoothly by a half cycle of sine curve, with curve equation is

$$P_{12}(x) = \frac{P_1 + P_2}{2} - \frac{P_1 - P_2}{2} \sin \frac{\pi(2x - x_1 - x_2)}{2(x_2 - x_1)} \qquad (12.15)$$

It can be easily proved that two endpoints of the curve are P_1 and P_2:

$$P_{12}(x_1) = \frac{P_1 + P_2}{2} - \frac{P_1 - P_2}{2} \sin \frac{\pi(2x_1 - x_1 - x_2)}{2(x_2 - x_1)} = P_1 \qquad (12.16)$$

$$P_{12}(x_2) = \frac{P_1 + P_2}{2} - \frac{P_1 - P_2}{2} \sin \frac{\pi(2x_2 - x_1 - x_2)}{2(x_2 - x_1)} = P_2 \qquad (12.17)$$

With regard to the curve to be connected at position of parameters P_1 and P_2, if slope is not 0, slope can be obtained in advance, and sine curve connects with it at the same slope. Long length is required to discuss the issue, so further study will not be made.

12.2.2 Transition of cylinder and airfoil

Now smooth transition of blade root cylinder and airfoil is regarded as an example to describe application of sine curve fairing method. Here blade root cylinder is regarded as a special airfoil to study how to transit with actual airfoil smoothly. Take cylinder as a special airfoil, function expression for round section of cylinder is required to transfer into the form that is equivalent to airfoil function expression, and compare parameters consistency at relevant position of these two function expressions, and carry out sine curve fairing for unequivalent expression. Taking previously discussed airfoil as an example still, discuss smooth transition between airfoil section and that blade root cylinder radius r_1 is 0.15 times. Set transition interval as [0.05, 0.25], $x_1=0.05$ inner side as cylinder, $x_2=0.25$ outer side as airfoil. Circular section profile equation is

$$z_{C1}^2 + (r_1 - y_C)^2 = r_1^2 \tag{12.18}$$

Convert into airfoil profile explicit expression, upward profile is

$$z_{Cu1} = \left[r_1^2 - (r_1 - y_C)^2\right]^{0.5} = y_C^{0.5}(d_1 - y_C)^{0.5} \tag{12.19}$$

$d_1=0.3$ represents cylinder diameter, y_C is taken from [0, 0.3] only, y_C can be taken from [0, 1], indicating cylinder diameter is 0.3 time of airfoil chord at the position $x=0.25$, this has changed chord length and embodied in change transition process, not necessarily embodied in chord function, therefore, "base" chord set at blade root zone and transition zone shall be identically equal to airfoil chord at the position that $x=0.25$.

Subscript u is used to represent upward profile, l represents downward profile; subscript 1 represents blade root cylinder, 2 represents airfoil, and subscript 12 represents transition zone. Convert Formula (12.19) into the way that is equivalent to airfoil function expression:

$$\begin{aligned} z_{Cu1} &= y_C^{0.5}(d_1 - y_C)^{0.5} \\ &= p_{u1} y_C^{a_{u1}}(1 - y_C)^{b_{u1}} + q_{u1} y_C^{0.5}(d_{u1} - y_C)^{c_{u1}} \end{aligned} \tag{12.20}$$

Airfoil upward profile is

$$\begin{aligned}z_{Cu2} &= 0.2y_C^1(1-y_C)^1 + 0.17y_C^{0.5}(1-y_C)^{1.5}\\ &= p_{u2}y_C^{a_{u2}}(1-y_C)^{b_{u2}} + q_{u2}y_C^{0.5}(d_{u2}-y_C)^{c_{u2}}\end{aligned} \quad (12.21)$$

Compare parameter values in these two profile functions are as shown in Table 12.1.

Table 12.1 Parameters comparison of blade root cylinder and airfoil function

Cylinder parameters		Airfoil parameters		Transition zone processing specification
Symbols	Value	Symbols	Value	
p_{u1}	0	p_{u2}	0.2	Unequal coefficient value, smooth transition is required
a_{u1}	1	a_{u2}	1	Equal exponent value, no processing is allowed
b_{u1}	1	b_{u2}	1	Equal exponent value, no processing is allowed
q_{u1}	1	q_{u2}	0.17	Unequal coefficient value, smooth transition is required
c_{u1}	0.5	c_{u2}	1.5	Unequal exponent value, smooth transition is required
d_{u1}	0.3	d_{u2}	1	Unequal constant value, smooth transition is required
p_{l1}	0	p_{l2}	0.02	Unequal coefficient value, smooth transition is required
a_{l1}	1	a_{l2}	1	Equal exponent value, no processing is allowed
b_{l1}	1	b_{l2}	1	Equal exponent value, no processing is allowed
q_{l1}	1	q_{l2}	0.37	Unequal coefficient value, smooth transition is required
c_{l1}	0.5	c_{l2}	1.5	Unequal exponent value, smooth transition is required
d_{l1}	0.3	d_{l2}	1	Unequal constant value, smooth transition is required

With regard to smooth transition for upward profile, it can be seen from Table12.1 that there are only four parameters that are required for smooth transition, with transition curves are

$$\begin{aligned}p_{u12} &= \frac{p_{u1}+p_{u2}}{2} - \frac{p_{u1}-p_{u2}}{2}\sin\frac{\pi(2x-x_1-x_2)}{2(x_2-x_1)}\\ &= \frac{0+0.2}{2} - \frac{0-0.2}{2}\sin\frac{\pi(2x-0.05-0.25)}{2(0.25-0.05)}\\ &= 0.1 + 0.1\sin(5\pi x - 0.75\pi)\end{aligned} \quad (12.22)$$

Similarly

$$q_{u12} = 0.585 - 0.415\sin(5\pi x - 0.75\pi) \quad (12.23)$$

$$c_{u12} = 1 + 0.5\sin\sin(5\pi x - 0.75\pi) \quad (12.24)$$

$$d_{u12} = 0.65 + 0.35\sin(5\pi x - 0.75\pi) \tag{12.25}$$

Then analyze smooth transition for downward profile. Downward profile for blade root cylinder is

$$z_{Cl1} = -y_C^{0.5}(d_1 - y_C)^{0.5} \tag{12.26}$$

Convert it into the way that is equivalent to airfoil function expression.

$$z_{Cl1} = p_{l1} y_C^{a_{l1}}(1-y_C)^{b_{l1}} - q_{l1} y_C^{0.5}(d_{l1} - y_C)^{c_{l1}} \tag{12.27}$$

Airfoil downward profile is

$$\begin{aligned} z_{Cl2} &= 0.02 y_C^1 (1-y_C)^1 - 0.37 y_C^{0.5}(1-y_C)^{1.5} \\ &= p_{l2} y_C^{a_{l2}}(1-y_C)^{b_{l2}} - q_{l2} y_C^{0.5}(d_{l2} - y_C)^{c_{l2}} \end{aligned} \tag{12.28}$$

Compare parameter values in these two profile functions, as shown in Table 12.1. There are also four parameters to be transited, with transition curves are

$$p_{l12} = 0.01 + 0.01\sin(5\pi x - 0.75\pi) \tag{12.29}$$

$$q_{l12} = 0.685 - 0.315\sin(5\pi x - 0.75\pi) \tag{12.30}$$

$$c_{l12} = 1 + 0.5\sin(5\pi x - 0.75\pi) \tag{12.31}$$

$$d_{l12} = 0.65 + 0.35\sin(5\pi x - 0.75\pi) \tag{12.32}$$

Substitute all transition parameters into respective airfoil profile expression, finally upward and downward profile expressions at smooth transition zone are

$$\begin{cases} z_{Cu} = \left[0.1 + 0.1\sin(5\pi x - 0.75\pi)\right] y_C (1 - y_C) \\ \qquad + \left[0.585 - 0.415\sin(5\pi x - 0.75\pi)\right] y_C^{0.5} \\ \qquad \left\{\left[0.65 + 0.35\sin(5\pi x - 0.75\pi)\right] - y_C\right\}^{[1+0.5\sin(5\pi x - 0.75\pi)]} \\ z_{Cl} = \left[0.01 + 0.01\sin(5\pi x - 0.75\pi)\right] y_C (1 - y_C) \\ \qquad - \left[0.685 - 0.315\sin(5\pi x - 0.75\pi)\right] y_C^{0.5} \\ \qquad \left\{\left[0.65 + 0.35\sin(5\pi x - 0.75\pi)\right] - y_C\right\}^{[1+0.5\sin(5\pi x - 0.75\pi)]} \end{cases} \tag{12.33}$$

In Formula (12.33), when $x=0.25$, indicating given airfoil, when $x=0.05$, it is degraded as circle, and when 0.05<x<0.25, indicating transitional graph from circle to given airfoil.

Transition method for different airfoils is similar to this method, no more detailed description will be made here.

12.3 Functional design steps and examples

12.3.1 Blade functional design steps

Blade functional design is mainly to determine blade function, and then generate blade stereo image by blade function and carry out calculation for blade performance. Specific steps are as follows.

1) Determine airfoil and calculate performance

Airfoil can be selected in airfoil library, and also generated by using function.

Lift drag ratio and tip speed ratio are most important factors to affect efficiency of wind turbine. The larger for airfoil lift drag ratio, the better. If higher tip speed ratio for wind turbine is expected, attack angle (optimal attack angle) that maximum lift drag ratio curve corresponded should not be too large, i.e., the product of optimum attack angle (degree) and tip speed ratio should not be greater than 37.7, otherwise it will produce negative twist at blade tip zone.

If airfoil is selected, it requires to approximate airfoil profile by function to obtain airfoil function expression, for the propose of constructing blade function and generating blade function graph.

After the airfoil is determined, performance data of lift and drag change along attack angle are required, and its function expression is obtained by function graph approximation method or regression analysis for wind turbine performance calculation.

2) Determination of design tip speed ratio

Design tip speed ratio is a fixed constant value given during design. Design tip speed ratio is neither too small nor too large. Too small ratio will cause excessive width and weight, increasing number of blade and too large gear speed, resulting in much difficulties during wind turbine production and installation. Design tip speed ratio also should not be too large, if so, blade speed is too fast and centrifugal force is large, blade may vibrate and the tip may whistle and will produce difficulty in starting.

Common range for design tip speed ratio of modern high speed wind turbine is 5-10, but the product of design tip speed ratio and optimum airfoil attack angle (degree) should not be greater than 37.7.

3) Determine number of blades

Number of blades for modern high-speed wind turbine is 2-4, and most are 3. Existing design and operating experience show that operating and power output stays relatively stable by using three blades[75].

4) Determine basic parameters for blade tip loss

Substitute axial speed induction factor under stable operation state (1/3) and above parameters into Prandtl blade tip loss formula to obtain blade tip loss correction factors, mean axial speed induction factor of wind rotor.

5) Calculate chord length and draw chord curve

Substitute above parameters into chord expression after blade tip loss correction to obtain chord function and generate function graph.

6) Simplify main part of curve chord into straight-line chord

Main part of curve chord is simplified into straight-line chord by tangent method, and tangent point can be set nearby the position $x=0.8$, and straight-line section can be set within [0.2,0.8]. Function expression for straight-line chord is established based on specific setting.

7) Establish twist function expression

Twist at blade tip part (outside of tangent point) is equal to difference of inflow angle and optimal attack angle of airfoil, and it can be calculated by twist formula.

Twist of straight-line chord should be re-calculated according to lift distribution condition. Substitute straight-line chord expression into lift coefficient formula to obtain lift distribution function along span; and then obtain attack angle distribution function along span by using airfoil lift and attack angle function relation; finally attack angle function subtracts inflow function to obtain twist distribution function along span.

8) Calculate performance of wind turbine

Substitute lift and drag coefficient distribution functions along span into practical wind turbine performance formula to obtain power, torque, lift and thrust coefficient values.

9) Optimize blade design

Repeat all above steps or part of steps to re-design, and compare wind turbine performance of various designs, select optimal result and complete optimal design steps. Optimal design only adjusts some parameters, for example adjusting tip speed ratio, moving tangent point position of straight-line chord etc.

10) Design blade root and transition zone

Blade root and transition zone are designed based on optimal design scheme. Blade root shape can be designed based on strength, installation difficulty, for example cylindrical blade root is used. Smooth transition can be made between blade root cylinder and airfoil at straight-line chord section with sine curve fairing method to obtain airfoil function expression at transition zone.

11) Construct blade function by segment

Divide blade into root, transition, straight-line chord and tip segment, and substitute airfoil function, chord function and twist function into blade function at various segments to obtain their function expressions.

12) Generate blade function graph

Substitute blade function expression at various segments into mathematical software (such as Mathematica) to generate function graph, and compose images at various images and upper and lower surfaces to obtain solid graph of the whole blade.

Subsequent steps include blade model print or making, numerical calculation or experimental study, and there will be no more detailed description here since it goes beyond the scope of this book.

12.3.2 Blade functional design example

Examples in Section 12.1 is still used to discuss specific process of blade functional design.

1) Determine airfoil and calculate performance

Set airfoil profile expression as

$$f_u(y_C) = 0.2y_C(1-y_C) + 0.17y_C^{0.5}(1-y_C)^{1.5} \tag{12.34}$$

$$f_l(y_C) = 0.02y_C(1-y_C) - 0.37y_C^{0.5}(1-y_C)^{1.5} \tag{12.35}$$

Airfoil along span always keeps unchanged.

Relation of airfoil lift, drag coefficient and attack angle under small attack angle state is

$$C_L = 0.12(\alpha + 3.6) \tag{12.36}$$

$$C_D = 0.012 + 1.052(\pi\alpha/180)^2 \tag{12.37}$$

The unit for attack angle in the formula is degree.

The relation between airfoil lift drag ratio and attack angle is

$$\zeta = \frac{C_L}{C_D} = \frac{0.12(\alpha + 3.6)}{0.012 + 1.052(\pi\alpha/180)^2} \tag{12.38}$$

Set $\partial\zeta/\partial\alpha = 0$ to obtain optimum attack angle of maximum lift drag ratio α_b as 3.5°, lift coefficient of the angle is 0.85, drag coefficient is 0.016.

2) Determination of design tip speed ratio

Design tip speed ratio $\lambda_t = 6$.

3) Determine number of blades

Number of blades for wind turbine is $B=3$.

4) Determine basic parameters for blade tip loss

Substitute axial speed induction factor under stable operation state ($a=1/3$) and above parameters into Prandtl blade tip loss formula to obtain blade tip loss correction factor.

$$f = \frac{2}{\pi}\arccos\left\{\exp\left[-\frac{3(1-x)\sqrt{1+81x^2}}{2x}\right]\right\} \tag{12.39}$$

In regard to wind turbine with limited blades, average axial speed induction factor by Formula (9.2) is

$$\bar{a} = \frac{1}{3} + \frac{1}{3}f - \frac{1}{3}\sqrt{1 - f + f^2} \tag{12.40}$$

Introduce the symbol g, set

$$g = 1 - \bar{a}/f \tag{12.41}$$

5) Calculate chord length and draw chord curve

Substitute basic parameters of wind turbine, blade tip correction factor and average axial speed induction factor under stable condition into the Formula (9.13) to obtain relative chord distribution function.

$$\frac{C}{R} = \frac{8\pi}{3}\frac{\bar{a}(1-\bar{a})x}{0.85\left(6x + \frac{\bar{a}g}{6xf}\right)\sqrt{\left(6x + \frac{\bar{a}g}{6xf}\right)^2 + g^2}} \tag{12.42}$$

Relative chord curve shape after correction of blade tip loss is shown in Figure 12.4.

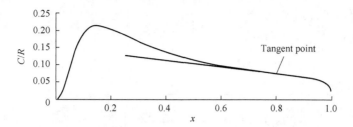

Figure 12.4 Chord curve with blade tip loss

6) Simplify main part of curve chord into straight-line chord

The chord curve should be simplified by using tangent method. Make tangent line at the position $x=0.78$, chord curve from outside of tangent point to blade tip keeps unchanged and straight from inside of tangent point to the position $x_2=0.25$.

Chord length at tangent point that $x=0.78$ is 0.074,9, slope is $-0.098,9$, thus tangent equation is

$$\frac{C}{R} = -0.098,9x + 0.152,0 \tag{12.43}$$

Tangent chord image is shown in the straight line in Figure 12.4.

7) Establish twist function expression

Substitute tangent Formula (12.43) into Formula (10.7) to back calculate lift coefficient, straight-line requires lift coefficient distribution at tangent zone $(0.25 \leqslant x \leqslant 0.78)$ as

$$C_L(x) = \frac{8\pi}{B} \frac{\bar{a}(1-\bar{a})x}{(-0.098,9x + 0.152,0)\left(\lambda_1 x + \frac{\bar{a}g}{\lambda_1 xf}\right)\sqrt{\left(\lambda_1 x + \frac{\bar{a}g}{\lambda_1 xf}\right)^2 + g^2}} \tag{12.44}$$

Attack angle distribution function along span is obtained by lift and attack angle of airfoil as

$$\alpha(x) = \frac{C_L(x)}{0.12} - 3.6$$

$$= \frac{200\pi}{3B} \frac{\bar{a}(1-\bar{a})x}{(-0.098,9x + 0.152,0)\left(\lambda_1 x + \frac{\bar{a}g}{\lambda_1 xf}\right)\sqrt{\left(\lambda_1 x + \frac{\bar{a}g}{\lambda_1 xf}\right)^2 + g^2}} - 3.6 \tag{12.45}$$

Twist at tangent zone is the difference between the inflow angle and above attack angle function.

$$\beta_1(x) = \varphi(x) - \alpha(x)$$
$$= \frac{180}{\pi} \arctan \frac{gf\lambda_t x}{f\lambda_t^2 x^2 + \bar{a}g}$$
$$- \frac{200\pi}{3B} \frac{\bar{a}(1-\bar{a})x}{(-0.098,9x + 0.152,0)\left(\lambda_t x + \frac{\bar{a}g}{\lambda_t xf}\right)\sqrt{\left(\lambda_t x + \frac{\bar{a}g}{\lambda_t xf}\right)^2 + g^2}} + 3.6$$
(12.46)

Chord at blade tip zone ($0.78 \leqslant x \leqslant 1$) is not changed, and adjustment for twist is not required, and twist at blade tip zone is the difference between the inflow angle and optimal attack angle:

$$\beta_2(x) = \varphi(x) - \alpha_b = \frac{180}{\pi} \arctan \frac{gf\lambda_t x}{f\lambda_t^2 x^2 + \bar{a}g} - 3.5 \qquad (12.47)$$

Twist parameters at various zones are shown in Table 12.2.

Table 12.2 Zone distribution for twist parameters

Spanwise position description	Location value	Relative twist value
Blade root zone	$x = 0 - 0.05$	0.435,3
Chord transition zone	$x = 0.05 - 0.25$	Determined by Formula (12.46)
Chord tangent zone	$x = 0.25 - 0.78$	Determined by Formula (12.46)
Blade tip zone	$x = 0.78 - 1$	Determined by Formula (12.47)

8) Calculate performance of wind turbine

Substitute lift and drag coefficient distribution function along span into practical wind turbine performance formula to calculate power, torque, lift and thrust coefficient value respectively.

a. Power performance calculation

Substitute basic parameters of wind turbine and lift and drag coefficient expressions into Formula (11.2), and power coefficient of straight-line chord section after numerical integration is obtained.

$$C_{P1} = \int_{0.25}^{0.78} 8\bar{a}(1-\bar{a})\lambda_t x^2 \left(\frac{g}{\lambda_t x + \frac{\bar{a}g}{\lambda_t xf}} - \frac{C_D}{C_L} \right) dx$$

$$= \int_{0.25}^{0.78} 8\bar{a}(1-\bar{a})\lambda_t x^2 \left(\frac{g}{\lambda_t x + \frac{\bar{a}g}{\lambda_t xf}} - \frac{0.012 + 1.052(\pi\alpha/180)^2}{0.12(\alpha+3.6)} \right) dx \quad (12.48)$$

$$= 0.2837$$

Attack angle at blade tip zone is optimal attack angle (3.5°), and power coefficient is

$$C_{P2} = \int_{0.8}^{1} 8\bar{a}(1-\bar{a})\lambda_t x^2 \left(\frac{g}{\lambda_t x + \frac{\bar{a}g}{\lambda_t xf}} - \frac{C_D}{C_L} \right) dx$$

$$= \int_{0.8}^{1} 8\bar{a}(1-\bar{a})\lambda_t x^2 \left(\frac{g}{\lambda_t x + \frac{\bar{a}g}{\lambda_t xf}} - \frac{0.012 + 1.052(3.5\pi/180)^2}{0.12(3.5+3.6)} \right) dx \quad (12.49)$$

$$= 0.163,2$$

Power coefficient for overall blade is

$$C_P = C_{P1} + C_{P2} = 0.283,7 + 0.163,2 = 0.446,9 \quad (12.50)$$

b. Torque performance calculation

Substitute basic parameters of wind turbine and lift and drag coefficient expressions into Formula (11.9), and torque coefficient of straight-line chord section after numerical integration is obtained.

$$C_{M1} = \int_{0.25}^{0.78} 8\bar{a}(1-\bar{a})x^2 \left(\frac{g}{\lambda_t x + \frac{\bar{a}g}{\lambda_t xf}} - \frac{C_D}{C_L} \right) dx$$

$$= \int_{0.25}^{0.78} 8\bar{a}(1-\bar{a})x^2 \left(\frac{g}{\lambda_t x + \frac{\bar{a}g}{\lambda_t xf}} - \frac{0.012 + 1.052(\pi\alpha/180)^2}{0.12(\alpha+3.6)} \right) dx \quad (12.51)$$

$$= 0.047,3$$

Attack angle at blade tip zone is optimal attack angle (3.5°), and torque coefficient is

$$C_{M2} = \int_{0.78}^{1} 8\bar{a}(1-\bar{a})x^2 \left(\frac{g}{\lambda_1 x + \frac{\bar{a}g}{\lambda_1 xf}} - \frac{C_D}{C_L} \right) dx$$

$$= \int_{0.78}^{1} 8\bar{a}(1-\bar{a})x^2 \left(\frac{g}{\lambda_1 x + \frac{\bar{a}g}{\lambda_1 xf}} - \frac{0.012 + 1.052(3.5\pi/180)^2}{0.12(3.5+3.6)} \right) dx \quad (12.52)$$

$$= 0.027,2$$

Torque coefficient for overall blade is

$$C_M = C_{M1} + C_{M2} = 0.047,3 + 0.027,2 = 0.074,5 \quad (12.53)$$

c. Lift performance calculation

Substitute basic parameters of wind turbine and lift and drag coefficient expressions into Formula (11.14), and lift coefficient of straight-line chord section after numerical integration is obtained.

$$C_{F1} = \int_{0.25}^{0.78} 8\bar{a}(1-\bar{a})x \left[\frac{g}{\lambda_1 x + \frac{\bar{a}g}{\lambda_1 xf}} - \frac{C_D}{C_L} \right] dx$$

$$= \int_{0.25}^{0.78} 8\bar{a}(1-\bar{a})x \left[\frac{g}{\lambda_1 x + \frac{\bar{a}g}{\lambda_1 xf}} - \frac{0.012 + 1.052(\pi\alpha/180)^2}{0.12(\alpha+3.6)} \right] dx \quad (12.54)$$

$$= 0.092,1$$

Attack angle at blade tip zone is optimal attack angle (3.5°), and lift coefficient is

$$C_{F2} = \int_{0.78}^{1} 8\bar{a}(1-\bar{a}) x \left[\frac{g}{\lambda_t x + \frac{\bar{a}g}{\lambda_t xf}} - \frac{C_D}{C_L} \right] dx$$

$$= \int_{0.78}^{1} 8\bar{a}(1-\bar{a}) x \left[\frac{g}{\lambda_t x + \frac{\bar{a}g}{\lambda_t xf}} - \frac{0.012 + 1.052(3.5\pi/180)^2}{0.12(3.5+3.6)} \right] dx \quad (12.55)$$

$$= 0.031,0$$

Lift coefficient for overall blade is

$$C_F = C_{F1} + C_{F2} = 0.092,1 + 0.031,0 = 0.123,1 \quad (12.56)$$

d. Thrust performance calculation

Substitute basic parameters of wind turbine into the formula, and thrust coefficient of the whole blade can be obtained after numerical integration.

$$C_T = \int_0^1 8\bar{a}(1-\bar{a}) x \, dx = 0.853,9 \quad (12.57)$$

9) Optimize blade design

Repeat all above steps or part of steps to re-design, and compare wind turbine performance of various designs and select optimal result. As an example, repeated design process is ignored here.

10) Design blade root and transition zone

Set blade root cylinder length as 0.05. Transition zone is set between blade root and blade, with interval of [0.05, 0.25]; blade chord length is 0.127,3 at x=0.25, take cylinder diameter as 0.3 time of the chord length, i.e., 0.038,2, main parameters are shown in Table 12.3.

Table 12.3 Value of chord curve at several key parts

Spanwise position description	Location value	Relative chord value
Outside of blade root cylinder	x=0.05	0.038,2
Inside at tangent zone	x=0.25	0.127,3
Chord tangent point	x=0.78	0.074,9
Blade tip zone	x=1	0
Blade root zone	x=0-0.05	0.038,2
Chord transition zone	x=0.05-0.25	Smooth transition is generated automatically
Chord tangent zone	x=0.25-0.78	$-0.098,9x+0.152,0$
Chord blade tip zone	x=0.78-1	Determined by Formula (12.42)

Blade root cylinder refers to the part from blade root to the position $x_1=0.05$, but smooth transition zone between x_1 and x_2 is no longer set with straight line, and sine curve fairing method is replaced to transit smoothly to airfoil from blade root cylinder.

Upward and downward airfoil expression at smooth transition zone is obtained by using sine curve fairing method and smooth transition for blade root cylinder and airfoil given in Section 12.2 as

$$\begin{cases} f_u(y_C) = \left[0.1+0.1\sin(5\pi x - 0.75\pi)\right]y_C(1-y_C) + \\ \qquad \left[0.585-0.415\sin(5\pi x - 0.75\pi)\right]y_C^{0.5} \\ \qquad \left\{\left[0.65+0.35\sin(5\pi x - 0.75\pi)\right]-y_C\right\}^{\left[1+0.5\sin(5\pi x - 0.75\pi)\right]} \\ f_l(y_C) = \left[0.01+0.01\sin(5\pi x - 0.75\pi)\right]y_C(1-y_C) - \\ \qquad \left[0.685-0.315\sin(5\pi x - 0.75\pi)\right]y_C^{0.5} \\ \qquad \left\{\left[0.65+0.35\sin(5\pi x - 0.75\pi)\right]-y_C\right\}^{\left[1+0.5\sin(5\pi x - 0.75\pi)\right]} \end{cases} \quad (12.58)$$

Power generated at smooth transition zone is small, and its twist can be calculated based on the formula, i.e., expand x interval to [0.05, 0.78].

11) Construct blade function

Substitute airfoil function, chord function and twist function into blade function at various segments to obtain their function expressions.

a. Construct segmented chord function

Chord function is divided into three segments: blade root and transition segment, straight-line chord segment and blade tip segment. With the chord value (0.123,7R) at the junction with straight-line chord segment for blade root and transition segment as the base, and actual chord is determined by transition function from airfoil to blade root cylinder. Chord function obtained finally is

$$C(x) = \begin{cases} 0.127,3R & (0 < x \leqslant 0.05) \\ (-0.098,9x+0.152)R & (0.05 < x \leqslant 0.78) \\ \dfrac{8\pi R}{3} \dfrac{\bar{a}(1-\bar{a})x}{0.85\left(6x+\dfrac{\bar{a}g}{6xf}\right)\sqrt{\left(6x+\dfrac{\bar{a}g}{6xf}\right)^2 + g^2}} & (x > 0.78) \end{cases} \quad (12.59)$$

b. Construct segmented twist function

Twist function is divided into three segments: blade root segment, transition and straight-line chord segment and blade tip segment. Twist value (0.435,3) at the junction is used for blade root segment. Twist function obtained finally is

$$\beta(x) = \begin{cases} 0.435,3 \ (x \leqslant 0.05) \\ -\dfrac{200\pi}{9} \dfrac{\bar{a}(1-\bar{a})x}{(-0.097,4x+0.150,6)\left(6x+\dfrac{\bar{a}g}{6xf}\right)\sqrt{\left(6x+\dfrac{\bar{a}g}{6xf}\right)^2+g^2}} \\ \quad + \dfrac{180}{\pi}\arctan\dfrac{6gfx}{36fx^2+\bar{a}g} + 3.6 \ (0.05 < x \leqslant 0.78) \\ \dfrac{180}{\pi}\arctan\dfrac{6gfx}{36fx^2+\bar{a}g} - 3.5 \ (x > 0.78) \end{cases} \quad (12.60)$$

c. Construct airfoil function

First, determine airfoil coefficient expression.

It can be known from Section 12.2, there are four coefficients changed for upward and downward profile during transition process from airfoil to cylinder, and after reaching blade root zone and straight-line chord segment, it keeps coefficient unchanged at the junction, therefore, each coefficient is segmented function.

Upward profile coefficient:

$$p_u = \begin{cases} 0 \ (x \leqslant 0.05) \\ 0.1 + 0.1\sin[\pi(-0.75+5x)] \ (0.05 < x \leqslant 0.25) \\ 0.2 \ (x > 0.25) \end{cases} \quad (12.61)$$

$$q_u = \begin{cases} 1 \ (x \leqslant 0.05) \\ 0.585 - 0.415\sin[\pi(-0.75+5x)] \ (0.05 < x \leqslant 0.25) \\ 0.17 \ (x > 0.25) \end{cases} \quad (12.62)$$

$$c_u = \begin{cases} 0.5 \ (x \leqslant 0.05) \\ 1 + 0.5\sin[\pi(-0.75+5x)] \ (0.05 < x \leqslant 0.25) \\ 1.5 \ (x > 0.25) \end{cases} \quad (12.63)$$

$$d_u = \begin{cases} 0.3 \ (x \leqslant 0.05) \\ 0.65 + 0.35\sin[\pi(-0.75 + 5x)] \ (0.05 < x \leqslant 0.25) \\ 1 \ (x > 0.25) \end{cases} \quad (12.64)$$

Downward profile coefficient:

$$p_l = \begin{cases} 0 \ (x \leqslant 0.05) \\ 0.01 + 0.01\sin[\pi(-0.75 + 5x)] \ (0.05 < x \leqslant 0.25) \\ 0.02 \ (x > 0.25) \end{cases} \quad (12.65)$$

$$q_l = \begin{cases} 1 \ (x \leqslant 0.05) \\ 0.685 - 0.315\sin[\pi(-0.75 + 5x)] \ (0.05 < x \leqslant 0.25) \\ 0.37 \ (x > 0.25) \end{cases} \quad (12.66)$$

$$c_l = \begin{cases} 0.5 \ (x \leqslant 0.05) \\ 1 + 0.5\sin[\pi(-0.75 + 5x)] \ (0.05 < x \leqslant 0.25) \\ 1.5 \ (x > 0.25) \end{cases} \quad (12.67)$$

$$d_l = \begin{cases} 0.3 \ (x \leqslant 0.05) \\ 0.65 + 0.35\sin[\pi(-0.75 + 5x)] \ (0.05 < x \leqslant 0.25) \\ 1 \ (x > 0.25) \end{cases} \quad (12.68)$$

Second, substitute all airfoil parameters transition curve into airfoil function $z_C = f(y_C)$ to replace original fixed parameters, so as to construct segmented airfoil function. Upward and downward function is

$$f(y_C) = \begin{cases} p_u y_C(1 - y_C) + q_u y_C^{0.5}(d_u - y_C)^{c_u} \\ p_l y_C(1 - y_C) - q_l y_C^{0.5}(d_l - y_C)^{c_l} \end{cases} \quad (12.69)$$

d. Construct blade function

Substitute above chord function, twist function and airfoil function into blade function parametric equation to construct blade function.

Upper blade surface parameter formula:

$$\begin{cases} x = x_R \\ y = \dfrac{C(x_R)}{R} y_C \cos\beta(x_R) - \dfrac{C(x_R)}{R}\left[p_u y_C(1-y_C) + q_u y_C^{0.5}(d_u - y_C)^{c_u} \right]\sin\beta(x_R) \\ z = \dfrac{C(x_R)}{R} y_C \sin\beta(x_R) + \dfrac{C(x_R)}{R}\left[p_u y_C(1-y_C) + q_u y_C^{0.5}(d_u - y_C)^{c_u} \right]\cos\beta(x_R) \end{cases}$$

(12.70)

Lower blade surface parameter formula:

$$\begin{cases} x = x_R \\ y = \dfrac{C(x_R)}{R} y_C \cos\beta(x_R) - \dfrac{C(x_R)}{R}\left[p_l y_C(1-y_C) - q_l y_C^{0.5}(d_l - y_C)^{c_l} \right]\sin\beta(x_R) \\ z = \dfrac{C(x_R)}{R} y_C \sin\beta(x_R) + \dfrac{C(x_R)}{R}\left[p_l y_C(1-y_C) - q_l y_C^{0.5}(d_l - y_C)^{c_l} \right]\cos\beta(x_R) \end{cases}$$

(12.71)

12) Generate blade function graph

Substitute upper blade surface parametric equation into mathematical software to facilitate to generate function graph, and obtain blade solid figure, and detailed description will be made by examples in the following chapters.

12.3.3 Blade image generated by software

Almost all mathematical softwares enable to generate three-dimension function graph, and we take Mathematica as an example, convert formula in above design examples into procedure code, so as to generate three-dimensional image of blade function. Formula code in Mathematica software is very close to common formula, and easy to understand, therefore, simple description is made only when code is given (see text part in code). Completed code to generate blade image (meaning for individual symbol changed slightly):

Blade tip loss correction function:

$$f = \dfrac{2}{\pi} \mathrm{ArcCos}\left[e^{-\dfrac{3(1-x)\sqrt{1+81x^2}}{2x}} \right]$$

Average axial speed induction factor:

$$a = \frac{1}{3} + \frac{1}{3}f - \frac{1}{3}\sqrt{1-f+f^2}$$

Intermediate variable g:

$$g = 1 - \frac{a}{f}$$

Segmented chord function:

$$\begin{array}{l} C := 0.1273 / ; x \leqslant 0.05 \\ C := -0.0989x + 0.152 / ; 0.05 < x \leqslant 0.78 \\ C := \dfrac{8\pi}{3} \dfrac{a(1-a)x}{0.85\left(6x + \dfrac{ag}{6xf}\right)\sqrt{\left(6x + \dfrac{ag}{6xf}\right)^2 + g^2}} / ; x > 0.78 \end{array}$$

Segmented twist function:

$$\begin{array}{l} \beta := 0.5056 / ; x \leqslant 0.05 \\ \beta := -\dfrac{200\pi}{9} \dfrac{a(1-a)x}{(-0.0974x+0.1506)\left(6x+\dfrac{ag}{6xf}\right)\sqrt{\left(6x+\dfrac{ag}{6xf}\right)^2+g^2}} + \dfrac{\pi}{180} \\ \quad\quad \text{ArcTan}\left[\dfrac{6gfx}{36fx^2+ag}\right] + \dfrac{3.6\pi}{180} / ; 0.05 < x \leqslant 0.78 \\ \beta := \text{ArcTan}\left[\dfrac{6gfx}{36fx^2+ag}\right] - \dfrac{3.5\pi}{180} / ; x > 0.78 \end{array}$$

Partial coefficient and exponent in airfoil upward profile formula:

$$\begin{array}{l} pu := 0 / ; x \leqslant 0.05 \\ pu := 0.1 + 0.1 \text{Sin}[\pi(-0.75+5x)] / ; 0.05 < x \leqslant 0.25 \\ pu := 0.2 / ; x > 0.25 \end{array}$$

$$\begin{array}{l} qu := 1 / ; x \leqslant 0.05 \\ qu := 0.585 - 0.415 \text{Sin}[\pi(-0.75+5x)] / ; 0.05 < x \leqslant 0.25 \\ qu := 0.17 / ; x > 0.25 \end{array}$$

$$\left.\begin{array}{l}\text{cu}:=0.5/;x\leqslant 0.05\\ \text{cu}:=1+0.5\text{Sin}[\pi(-0.75+5x)]/;0.05<x\leqslant 0.25\\ \text{cu}:=1.5/;x>0.25\end{array}\right]$$

$$\left.\begin{array}{l}\text{du}:=0.3/;x\leqslant 0.05\\ \text{du}:=0.65+0.35\text{Sin}[\pi(-0.75+5x)]/;0.05<x\leqslant 0.25\\ \text{du}:=1/;x>0.25\end{array}\right]$$

Partial coefficient and exponent in airfoil downward profile formula:

$$\left.\begin{array}{l}\text{pl}:=0/;x\leqslant 0.05\\ \text{pl}:=0.01+0.01\text{Sin}[\pi(-0.75+5x)]/;0.05<x\leqslant 0.25\\ \text{pl}:=0.02/;x>0.25\end{array}\right]$$

$$\left.\begin{array}{l}\text{ql}:=1/;x\leqslant 0.05\\ \text{ql}:=0.685-0.315\text{Sin}[\pi(-0.75+5x)]/;0.05<x\leqslant 0.25\\ \text{ql}:=0.37/;x>0.25\end{array}\right]$$

$$\left.\begin{array}{l}\text{cl}:=0.5/;x\leqslant 0.05\\ \text{cl}:=1+0.5\text{Sin}[\pi(-0.75+5x)]/;0.05<x\leqslant 0.25\\ \text{cl}:=1.5/;x>0.25\end{array}\right]$$

$$\left.\begin{array}{l}\text{dl}:=0.3/;x\leqslant 0.05\\ \text{dl}:=0.65+0.35\text{Sin}[\pi(-0.75+5x)]/;0.05<x\leqslant 0.25\\ \text{dl}:=1/;x>0.25\end{array}\right]$$

Airfoil upward profile function:

$$\text{fu} = \text{pu}\cdot y(1-y) + \text{qu}\cdot y^{0.5}(\text{du}-y)^{\text{cu}}$$

Airfoil downward profile function:

$$\text{fl} = \text{pl}\cdot y(1-y) - \text{ql}\cdot y^{0.5}(\text{dl}-y)^{\text{cl}}$$

x direction variable for upper blade surface parametric equation:

$$\text{xu} = x$$

y direction variable for upper blade surface parametric equation:

$$\text{yu} = C\cdot y\cdot \text{Cos}[\beta] - C\cdot \text{fu}\cdot \text{Sin}[\beta]$$

z direction variable for upper blade surface parametric equation:

$$zu = C \cdot y \cdot Sin[\beta] + C \cdot fu \cdot Cos[\beta]$$

x direction variable for lower blade surface parametric equation:

$$xl = x$$

y direction variable for lower blade surface parametric equation:

$$yl = C \cdot y \cdot Cos[\beta] - C \cdot fl \cdot Sin[\beta]$$

z direction variable for lower blade surface parametric equation:

$$zl = C \cdot y \cdot Sin[\beta] + C \cdot fl \cdot Cos[\beta]$$

Generate blade stereo image by parameterization drawing method:

```
ParametricPlot3D[{{xu,yu,zu},{xl,yl,zl}},{x,0,1},{y,0,1}]
```

Blade stereo image almost may generate instantaneously while running the program code, as shown in Figure 12.5.

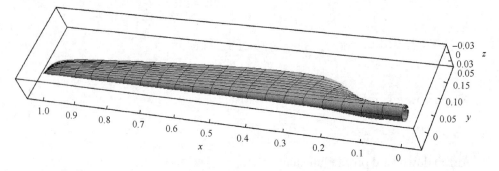

Figure 12.5　Three-dimensional blade image generated by software

Change the last code into

```
ParametricPlot3D[{{xu,yu,zu},{xl,yl,zl}},{x,0,1},{y,0,1},PlotStyle→None]
```

can generate line frame drawing, as shown in Figure 12.6.

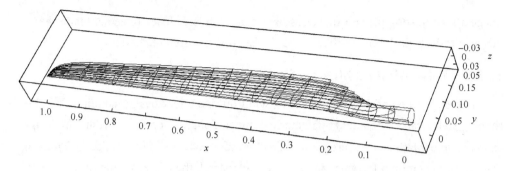

Figure 12.6 Three-dimensional blade line frame drawing generated by software

If translating z axis of airfoil to middle part of chord (x axis will move), so blade set can be set at middle part of airfoil, three-dimensional image is shown in Figure 12.7.

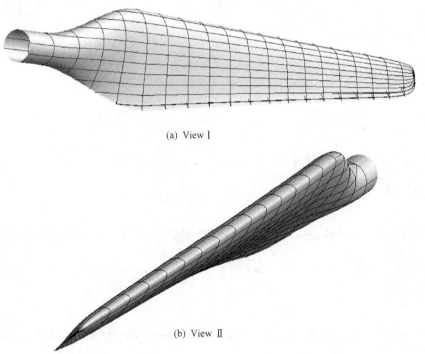

(a) View I

(b) View II

Figure 12.7 Blade solid drawing (blade root is located middle part of airfoil)

If blade root is set aerodynamic center position of airfoil, it will be realized by this method.

It can be seen from Figures 12.5-12.7, with consideration of actual factors, blade image after adjusting the sub-functions has been relatively closed to actual blade shape, this also reflects preliminary effect of blade functional design method.

12.3.4 Solve wind turbine performance by software

Power, torque and lift coefficient of wind turbine under design condition (stable running state) can be obtained by numerical integration, now convert formula in design example into Mathematica program code. Formula code in Mathematica software is very close to common formula, and easy to understand, therefore, simple description is made only when code is given (see text part in code). Completed code to calculate wind turbine performance (meaning for individual symbol changed slightly):

Blade tip loss function:

$$f = \frac{2}{\pi} \text{ArcCos}\left[e^{-\frac{3(1-x)\sqrt{1+81x^2}}{2x}} \right]$$

Average axial speed induction factor:

$$a = \frac{1}{3} + \frac{1}{3}f - \frac{1}{3}\sqrt{1-f+f^2}$$

Intermediate variable g:

$$g = 1 - \frac{a}{f}$$

Lift coefficient distribution function at straight-line chord segment along span:

$$CL = 0.12 \frac{\frac{200\pi}{9} a(1-a)x}{(-0.0974x+0.1506)\left(6x+\frac{ag}{6xf}\right)\sqrt{\left(6x+\frac{ag}{6xf}\right)^2+g^2}}$$

Drag coefficient distribution function at straight-line chord segment along span:

$$CD = 0.012 + 1.052\left(\left(\pi \frac{\frac{200\pi}{9} a(1-a)x}{(-0.0974x+0.1506)\left(6x+\frac{ag}{6xf}\right)\sqrt{\left(6x+\frac{ag}{6xf}\right)^2+g^2}} - 3.6\right)/180\right)^2$$

Power coefficient integrand at straight-line chord segment:

$$DCP1 = 48a(1-a)x^2\left(\frac{g}{6x+\frac{ag}{6xf}} - \frac{CD}{CL}\right)$$

Integrand of power coefficient at blade tip segment:

$$DCP2 = 48a(1-a)x^2\left(\frac{g}{6x+\frac{ag}{6xf}} - \frac{0.016}{0.85}\right)$$

Numerical integration solves total power coefficient:

`NIntegrate[DCP1,{x,0.25,0.78}]+NIntegrate[DCP2,{x,0.78,1}]`

Torque coefficient integrand at straight-line chord segment:

$$DCM1 = 48a(1-a)x\left(\frac{g}{6x+\frac{ag}{6xf}} - \frac{CD}{CL}\right)$$

Integrand of torque coefficient at blade tip segment:

$$DCM2 = 48a(1-a)x\left(\frac{g}{6x+\frac{ag}{6xf}} - \frac{0.016}{0.85}\right)$$

Numerical integration solves total torque coefficient:

`NIntegrate[DCM1,{x,0.25,0.78}]+NIntegrate[DCM2,{x,0.78,1}]`

Lift coefficient integrand at straight-line chord segment:

$$DCF1 = 8a(1-a)x\left(\frac{g}{6x+\frac{ag}{6xf}} - \frac{CD}{CL}\right)$$

Integrand of lift coefficient at blade tip segment:

$$DCF2 = 8a(1-a)x\left(\frac{g}{6x+\frac{ag}{6xf}} - \frac{0.016}{0.85}\right)$$

Numerical integration solves total lift coefficient:

```
NIntegrate[DCF1,{x,0.25,0.78}]+NIntegrate[DCF2,{x,0.78,1}]
```

Run the program code in Mathematica software, power, torque and lift performance of wind turbine will be calculated instantaneously. This calculation will not depend on blade shape, it can be completed before complex blade shape design, and is good for optimal design and improving design efficiency.

12.4　Summary of this Chapter

This chapter established general expression of blade mathematical model (blade curved surface function) by dimensional consistency rotation and rotation of twist on airfoil. It discussed smooth transition between blade root cylinder and airfoil in examples by sine curve fairing method. Studies have shown that blade mathematical model is composed of airfoil function, chord function and twist function, and can be expressed as explicit parameter equation set with two parameters (span position and chord position), and all variables in the equation have definite geometrical significance, they can be adjusted, assigned and modified easily and quickly.

This chapter also proposed specific steps and examples for blade functional design, gave method and program code to generate blade three-dimensional image by blade function with Mathematica software, as well as that to solve the wind turbine performance by using Mathematica software. Examples and studies have shown that blade functional design enables to realize three-dimension design of practical blade structure, and calculate its performance by analytical method, and it is a fully new method for blade design.

Chapter 13　Steps of Designing Solid Model of Blade

In the manufacturing process of wind turbine blade, both the design and processing of blade mold cavity and the design of the ply of blade made of composite materials require the use of three-dimensional solid model or surface model of blade. Therefore, the conversion of surface equation of blade to corresponding three-dimensional solid model or surface model is essential for the transformation of blade design into product.

13.1　Design idea

According to the idea of creating a solid model with the surface equation of blade, the common process is: generate corresponding section profiles based on the given intercept, and then loft multiple section profiles to obtain surface of blade, and finally repair the surface. Therefore, to obtain the solid model of blade, a technical route which is based on three-dimension design software SolidWorks is adopted and described as follows:

(1) Do secondary development based on SolidWorks VBA, calculate each section curve, and insert the curves into the modeling environment in SolidWorks automatically;

(2) Use the available surface molding tools to loft the section curves to generate blade surface in the software SolidWorks;

(3) Do surface repair, splicing, stitching and the like on the surface model of blade to complete blade modeling.

13.2　Insert section curves

The method of inserting section curves of blade is discussed below by following the above technical route with an example used in Chapter 12.

13.2.1　Determine the surface equation

Define the coordinate system of blade, as shown in Figure 13.1: y axis represents the direction of airfoil chord line (the leading edge-trailing edge direction is positive), z axis is aligned with the direction of incoming air, and x axis is perpendicular with Oyz plane (the root-tip direction is positive).

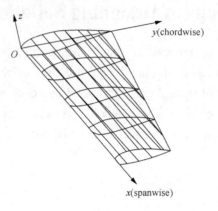

Figure 13.1 Example of coordinate system of blade

The known surface formula of blade is as follows:

$$\begin{cases} y = \dfrac{C(x)}{R} y_C \cos\beta(x) - \dfrac{C(x)}{R} f(y_C) \sin\beta(x) \\ z = \dfrac{C(x)}{R} y_C \sin\beta(x) + \dfrac{C(x)}{R} f(y_C) \cos\beta(x) \end{cases} \quad (13.1)$$

where R is the length of blade; x is the spanwise relative position of section curve which falls within a change interval of [0, 1]; y_C is the chordwise relative position of airfoil which falls within a change interval of [0, 1].

In Formula (13.1), chord function $C(x)$ is a piecewise function which is defined as

$$C(x) = \begin{cases} 0.127,3R & (0 < x \leqslant 0.05) \\ (-0.098,9x + 0.152)R & (0.05 < x \leqslant 0.78) \\ \dfrac{8\pi R}{3} \dfrac{\bar{a}(1-\bar{a})x}{0.85\left(6x + \dfrac{\bar{a}g}{6xf}\right)\sqrt{\left(6x + \dfrac{\bar{a}g}{6xf}\right)^2 + g^2}} & (x > 0.78) \end{cases} \quad (13.2)$$

The tip loss correction factor in Formula (13.2):

$$f = \frac{2}{\pi}\arccos\left[e^{-\frac{3(1-x)\sqrt{1+81x^2}}{2x}}\right]$$

The average axial induced velocity factor.

$$\bar{a} = \frac{1}{3} + \frac{1}{3}f - \frac{1}{3}\sqrt{1-f+f^2} \tag{13.3}$$

Intermediate variable

$$g = 1 - \frac{a}{f} \tag{13.4}$$

The blade twist function $\beta(x)$ is also a piecewise function which is expressed as follows:

$$\beta(x) = \begin{cases} 0.435,3 \ (x \leqslant 0.05) \\ -\dfrac{200\pi}{9} \dfrac{\bar{a}(1-\bar{a})x}{(-0.097,4x+0.150,6)\left(6x+\dfrac{\bar{a}g}{6xf}\right)\sqrt{\left(6x+\dfrac{\bar{a}g}{6xf}\right)^2+g^2}} \\ +\dfrac{180}{\pi}\left(\arctan\dfrac{6gfx}{36fx^2+\bar{a}g}+3.6\right) (0.05 < x \leqslant 0.78) \\ \dfrac{180}{\pi}\left(\arctan\dfrac{6gfx}{36fx^2+\bar{a}g}-3.5\right)(x>0.78) \end{cases} \tag{13.5}$$

The upper profile and lower profile function $f(y_C)$ of the airfoil in blade formula is defined as

$$f(y_C) = \begin{cases} p_u y_C(1-y_C) + q_u y_C^{0.5}(d_u - y_C)^{c_u} \\ p_l y_C(1-y_C) - q_l y_C^{0.5}(d_l - y_C)^{c_l} \end{cases} \tag{13.6}$$

Where the upper profile coefficient:

$$p_u = \begin{cases} 0 \ (x \leqslant 0.05) \\ 0.1 + 0.1\sin[\pi(-0.75+5x)] \ (0.05 < x \leqslant 0.25) \\ 0.2 \ (x > 0.25) \end{cases}$$

$$p_u = \begin{cases} 0 & (x \leqslant 0.05) \\ 0.1 + 0.1\sin[\pi(-0.75 + 5x)] & (0.05 < x \leqslant 0.25) \\ 0.2 & (x > 0.25) \end{cases}$$

$$c_u = \begin{cases} 0.5 & (x \leqslant 0.05) \\ 1 + 0.5\sin[\pi(-0.75 + 5x)] & (0.05 < x \leqslant 0.25) \\ 1.5 & (x > 0.25) \end{cases} \quad (13.7)$$

$$d_u = \begin{cases} 0.3 & (x \leqslant 0.05) \\ 0.65 + 0.35\sin[\pi(-0.75 + 5x)] & (0.05 < x \leqslant 0.25) \\ 1 & (x > 0.25) \end{cases}$$

The lower profile coefficient:

$$p_l = \begin{cases} 0 & (x \leqslant 0.05) \\ 0.01 + 0.01\sin[\pi(-0.75 + 5x)] & (0.05 < x \leqslant 0.25) \\ 0.02 & (x > 0.25) \end{cases}$$

$$q_l = \begin{cases} 1 & (x \leqslant 0.05) \\ 0.685 - 0.315\sin[\pi(-0.75 + 5x)] & (0.05 < x \leqslant 0.25) \\ 0.37 & (x > 0.25) \end{cases}$$

$$c_l = \begin{cases} 0.5 & (x \leqslant 0.05) \\ 1 + 0.5\sin[\pi(-0.75 + 5x)] & (0.05 < x \leqslant 0.25) \\ 1.5 & (x > 0.25) \end{cases} \quad (13.8)$$

$$d_l = \begin{cases} 0.3 & (x \leqslant 0.05) \\ 0.65 + 0.35\sin[\pi(-0.75 + 5x)] & (0.05 < x \leqslant 0.25) \\ 1 & (x > 0.25) \end{cases}$$

The surface formula of blade has been fully determined so far.

According to the modeling idea, it is possible to obtain the section curve in position x with Formulas (13.7) and (13.8) if different spanwise positions are given (namely, let x=0.1, 0.2, 0.3, 0.4,⋯, 1). Further lofting of these section curves will deliver the surface of blade.

Further observe this surface formula. Attention shall be paid to the following:

(1) By definition of the function $C(x)$, the value of x cannot be 0, or there will be an error as the divisor is zero. In other words, the section when $x=0$ cannot be obtained through program-based calculations but manual addition;

(2) By definition of f and g, the value of x cannot be 1. If $x=1, f=0$, g cannot be calculated. Therefore, it is impossible to obtain the data about the tip of blade through program-based calculations but manual termination in the last step;

(3) By definition of $f(y_C)$, with respect to a specific section curve, the value of y_C cannot be larger than d_u (upper profile) or d_l (lower profile);

(4) The surface at the root of blade is actually a cylindrical surface within the interval of $x \leqslant 0.05$.

13.2.2 Insert section curves in SolidWorks

Based on the shown foramulas and some simple properties, it is possible to calculate the coordinates of each point on corresponding section curves and insert corresponding curves in the modeling environment in SolidWorks. An example of inserting curve is as shown in Figure 13.2.

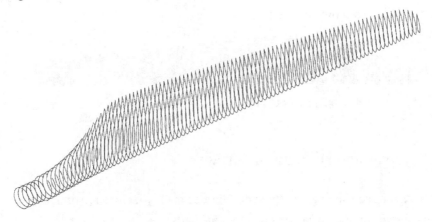

Figure 13.2 Example of inserting section curves of blade

Three issues shall be noticed.

(1) Number of sections: the larger the number of sections is, the higher the accuracy may be, but the time spent for calculations and future modeling and processing will be inevitably and greatly prolonged, so the number of sections shall fall within a reasonable range.

(2) Number of segments on each section curve: the equivalent of how many

straight lines are used to approximate the actual curve. The more the number of segments is, the closer the obtained curve is to actual curve. But it must be taken into account that the curves inserted in SolidWorks are subject to automatic fitting to ensure continuity of the curves, so even a few points will deliver a smooth curve.

(3) Each section curve is divided into upper curve and lower curve which are inserted separately, which is more helpful for obtaining a high quality surface in subsequent steps.

13.3 Do lofting to obtain surface

Most three-dimensional design software provides lofting function. Taking SolidWorks as example, it is simply needed to check the curves to be lofted in order to preview the shape of lofted surface.

It is possible that some curves are not closed after inserting the completed section curves of blade; therefore, it is not possible to obtain the entire blade through lofting, and instead the surface has to be lofted segmentally first and then spliced. The optional methods include: segmental lofting, fragmentary lofting, and then spliced to get a complete surface, as shown in Figure 13.3.

Figure 13.3 Example of segmental lofting of surface

13.4 Repair and splicing of surface

The repair and splicing on segmented surface mainly include repair at the opening, repair on surface of tip and splicing between the parts.

1) Repair at the opening on surface of blade

The repair at the opening on blade may be realized through the surface lofting function. The lofting stops at the sidelines on both sides of the opening.

2) Repair on surface of tip

For surfaces of the tip, it is optional to create a space curve which is deemed as the edge, and then do lofting with the sideline and edge of the surface of blade to obtain desired surface of tip. The curve of edge has to be tangent to the sidelines on both sides of the end of blade, and meanwhile the height of edge has to be equal to a section intercept.

3) Splicing of three segments of the surface of blade

The modeling at shank of the blade only requires lofting the surface. For relatively complex splicing positions, splicing may be realized through fragmentary lofting. Finally, stitch all segmental and fragmentary surfaces to obtain the solid model of the blade, as shown in Figure 13.4.

Figure 13.4　Example of solid model of blade

13.5　Summary of this chapter

By referring to an example, this chapter provides the steps of designing a solid model of blade and also covers discussion about the problems that are likely to be encountered during blade modeling and corresponding solutions. The design steps mainly consist of: create the section profile curves of blade, insert the curves, do lofting of the inserted curves, generate blade surface, doing repair, splicing and stitching of defective areas on surface, and finally obtain the solid model of blade.

References

[1] van Kuik G A M. The Lanchester-Betz-Joukowsky limit[J]. Wind Energy, 2007, (10): 289-291.

[2] Johansen J, Madsen H A, Gaunaa M, et al. Design of a wind turbine rotor for maximum aerodynamic efficiency[J]. Wind Energy, 2009, 12(3): 261-273.

[3] Ray T, Tsai H M. Swarm algorithm for single and multiobjective airfoil design optimization[J]. American Institute of Aeronautics and Astronautics (AIAA), 2004, 42(2): 366-373.

[4] David W Z, Timothy M L, Laslo D. Improvements to a Newton-Krylov adjoint algorithm for aerodynamic optimization[R]. AIAA 2005-4857, 17th AIAA Computational Fluid Dynamics Conference, 2005.

[5] Hicks R M, Henne A P. Wing design by numerical optimization[J]. Aircraft, 1978, 15(7): 407-413.

[6] Chen J, Zhang S Q, Eecen P J, et al. Wind turbine airfoil parametric expression and convergence characteristics[J]. Chinese Journal of Mechanical Engineering, 2010, 46(10): 132-138.

[7] Chen J, Wang X D, Shen W Z, et al. Blades shape optimization design for wind turbine[J]. Chinese Journal of Mechanical Engineering, 2010, 46(3): 131-134.

[8] Gur O, Rosen A. Optimal design of horizontal axis wind turbine blades[C]. 2008 Proceedings of the 9th Biennial Conference on Engineering Systems Design and Analysis, 2009, 3: 99-109.

[9] Chen J, Zhang S Q, Lu Q F, et al. The application of functional analysis idea in new wind turbine blades design[J]. Journal of Chongqing University, 2011, (7): 14-19.

[10] Mejía J M, Chejne F, Smith R, et al. Simulation of wind energy output at Guajira, Colombia[J]. Renewable Energy, 2006, 31(3): 383-399.

[11] Maalawi K Y, Badr M A. A practical approach for selecting optimum wind rotors[J]. Renewable Energy, 2003, 28(5): 803-822.

[12] Capuzzi M, Pirrera A, Weaver P M. A novel adaptive blade concept for large-scale wind turbines. Part II: Structural design and power performance[J]. Energy, 2014, (73): 25-32.

[13] Chen J H, Wang T G. Aeroelastic response calculation of wind turbine blades under yawing condition[J]. Journal of Nanjing University of Aeronautics and Astronautics, 2011, 43(5): 629-634.

[14] Wilson R E, Lissaman P B S, Walker S N. Aerodynamic performance of wind turbines[R]. Corvallis: Oregon State University, 1976.

[15] Zhao X, Xiao J, Xi D. Comparison of horizontal axis wind turbine under Wilson method and compound optimization method[J]. Journal of Northwestern Polytechnical University, 2008, 26(6): 693-697.

[16] Li G N, Yang F Z, Du B S, et al. MATLAB and Pro/E based wind turbine rotor design and modelling[J]. Mechanical Design, 2009, 26(6): 3-6.

[17] Wang F, Wang M T, Li X T, et al. Research status and prospect of horizontal axis wind turbine blades[J]. Mechanical Design, 2015, 11(32): 1-7.

[18] Jiang H B, Zhao Y P, Cheng Z Q. Analytical method and its position and role in complex problem study[J]. Chinese Journal of Systems Science, 2014, 22(1): 56-59.

[19] Colwell R R. Complexity and connectivity: A new cartography for science and engineering[C]. Remarks from the American Geophysical Union's fall meeting. San Francisco, 1999.

[20] Korn G A, Wait J V. Digital Continuous System Simulation[M]. Prentice-Hall, Inc., Englewood Cliffs, New Jersey, 1978.
[21] Xiao T Y. Is simulation based on model experiment?[J]. Journal of System Simulation, 2009, 21(22): 7368-7371.
[22] Burton T, Jenkins N, Sharpe D, et al. Wind Energy Handbook(2nd ed)[M]. Chichester: Wiley, 2011.
[23] Sumner J, Masson C. Influence of atmospheric stability on wind turbine power performance curves[J]. Wind Energy Engineering, 2006, 128: 531-537.
[24] Liao M F, Song W P, Wang S J, et al. Wind Turbine Design Theory and Structural Dynamics[M]. Xi'an: Northwestern Polytechnical University Press, 2014.
[25] Grigorescu D. Wind Turbine Theory and Design[M]. Translated by Shi P F. Beijing: China Machine Press, 1987.
[26] Zhao D P, Xu B Q. Wind Turbine Design Theory and Methods[M]. Beijing: Peking University Press, 2012.
[27] Jiang H B, Cao S L, Li Y R. Horizontal axis wind turbine blades twist and idea chord distribution[J].ActaEnergiae Solaris Sinica, 2013, 33(1): 1-6.
[28] He D X. Wind Engineering and Industrial Aerodynamics[M]. Beijing: National Defense Industry Press, 2006.
[29] Glauert H. Airplane propellers//Durand W F. Aerodynamic Theory[M]. New York: Dover, 1963.
[30] Hansen M O L. Aerodynamics of Wind Turbines(Edition II)[M]. Beijing: China Electric Power Press, 2009.
[31] Dong L, Liao M F, Ding Y W. Wind turbine blade aerodynamic design and unbalance loading calculation[J]. Acta Energiae Solaris Sinica, 2009, 30(1): 122-127.
[32] Jiang H B, Cao S L, Yang P. Power limit for horizontal axis wind turbine[J]. Chinese Journal of Mechanical Engineering, 2011, 47(10): 113-118.
[33] Wu S Q, Zhao D P. Aerodynamics of Wind Turbines[M]. Beijing: Peking University Press, 2011.
[34] Jiang H B, Cheng Z Q, Zhao Y P. Torque limit of horizontal axis wind turbine[C]. Proceedings of 2012 International Conference on Mechanical Engineering and Material Science. Shanghai, 2012: 148-151.
[35] Jiang H B, Zhao Y P, Cheng Z Q. Lift limit of horizontal axis wind turbine[J]. Advanced Materials Research, 2014, 1070-1072:1869-1873.
[36] Griffiths R T. The effect of airfoil characteristics on windmill performance[J]. Aeronautical Journal, 1977, 81(7): 322-326.
[37] Hassanein A, El-Banna H, Abdel-Rahman M. Effectiveness of airfoil aerodynamic characteristics on wind turbine design performance[C]. Proceedings of the Seventh International Conference on Energy and Environment. Cairo, 2000: 525-537.
[38] Kong C, Kim T, Han D, et al. Investigation of fatigue life for a medium scale composite wind turbine blade[J]. International Journal of Fatigue, 2006, 28(10): 1382-1388.
[39] Lobitz D W. Aeroelastic stability predictions for a MW-sized blade[J]. Wind Energy, 2004, 7: 211-224.
[40] Chaviaropoulos P K. Flap/lead-lag aeroelastic stability of wind turbine blades[J]. Wind Energy, 2001, 4: 183-200.
[41] Kottapalli S B R, Friedmann P P. Aeroelastic stability and response of horizontal axis wind turbine blades[J]. AIAA Journal, 1979, 17(12): 1381-1388.
[42] Yu H N, Zhu F L, Liu Y. Overview of wind turbine aeroelasticstability[J]. Mechanical Design, 2008, 25(6): 1-3.
[43] Chen X B, Li J, Chen J Y. Rotary wind turbine blade dynamic characteristics analysis considering centrifugal stealing effect[J]. Journal of Earthquake Engineering and Vibration, 2009, 29(1): 117-122.
[44] Liu X, Li G Q, Chen Y, et al. Horizontal axis wind turbine blade response analysis[J]. Chinese Journal of Mechanical Engineering, 2010, 46(12): 128-134.

[45] Si H Q, Wang T G, Wu X J. Impact study on parameters on wind turbine aerodynamic noise[J]. Acta Aerodynamica Sinica, 2014, 32(1): 131-135.

[46] Taehyung K I M, Seungmin L E E, Hogeon K I M, et al. Design of low noise airfoil with high aerodynamic performance for use on small wind turbines[J]. Science China(Technological Sciences), 2010, 53(1): 75-79.

[47] Xu W D. Fluid Mechanics[M]. Beijing: National Defend Industry Press, 1979.

[48] Jiang H B, Cao S L, Cheng Z Q. Calculation of panel high attack angle streaming lift and resistance coefficient[J]. Chinese Journal of Applied Mechanics, 2011, 28(5): 518-520.

[49] Mechanical Engineering Handbook, Electrical Engineering Handbook Editorial Committee. Mechanical Engineering Handbook Edition II (basic theory volume Chapter 7 Fluid mechanics) [M]. Beijing: China Machine Press, 1997.

[50] Hoerner S F. Fluid-Dynamic Drag, Hoerner Fluid Dynamics[M]. Bricktown, New Jersey, 1965.

[51] Viterna L A, Janetzke D C. Theoretical and experimental power from large horizontal-axis wind turbines[C]. Fifth Biennal Wind Energy Conference and Workshop. Washington D C, 1981.

[52] Ostowari C, Naik D. Post stall studies of untwisted varying aspect ratio blades with an NACA 4415 airfoil section-part I[J]. Wind Engineering, 1984, 8(3): 176-194.

[53] Zhang Z W. Experimental study on wind turbine two-dimensional airfoil high attack angle[J]. Acta Energiae Solaris Sinica, 1988, 9(1): 74-79.

[54] Sheldahl R E, Klimas P C. Aerodynamic characteristics of seven symmetrical airfoil sections through 180-Degree angle of attack for use in aerodynamic analysis of vertical axis wind turbines[R]. Sandia National Laboratories, Report SAND80-2114, 1981.

[55] Jiang H B, Li Y R, Cheng Z Q. Relations of lift and drag coefficients of flow around flat plate[J]. Applied Mechanics and Materials, 2014, 518: 161-164.

[56] Wright A K, Wood D H. The starting and low wind speed behaviour of a small horizontal axis wind turbine[J]. Journal of Wind Engineering and Industrial Aerodynamics, 2004, (92): 1265-1279.

[57] Ji R J. Wind turbine blade design method development[J]. China Electric Power Education, 2006, (1): 129-131.

[58] Dong Z N, Zhang Z X. Non-viscous Fluid Mechanics[M]. Beijing: Tsinghua University Press, 2003.

[59] Jiang H B, Zhao Y P. Mean line-thickness function-based airfoil shape analytic construction method[J]. Graphics Journal, 2013, 34(1): 50-54.

[60] Jiang H B, Cheng Z Q, Li Y R, et al. Airfoil indicated by analytical function and generation method[J]. Patent Gazette, 2013, 29(24): 66-73.

[61] Casey M. A computational geometry for the blades and internal flow channels of centrifugal compressors[J]. ASME Journal of Engineering for Power, 1983, 105(4): 288-295.

[62] Li S Z, Zhao F, Yang L. CFD-based airfoil hydrodynamic performance multi-objective optimal design[J]. Journal of Ship Mechanics, 2010, 14(11): 1241-1248.

[63] Chen J, Wang Q. Wind Turbine Airfoil and Blade Optimal Design Theory[M]. Beijing: Science Press, 2013.

[64] Cai X, Fan P, Zhu J, et al. Wind Power Generation Blade[M]. Beijing: China Water & Power Press, 2014.

[65] Jiang H B, Cheng Z Q, Zhao Y P. Function airfoil and its pressure distribution and lift coefficient calculation[C]. Proceedings of 3rd International Conference on Mechanical Science and Engineering. Hong Kong, 2013.

[66] Han J R. Approximate solution for any airfoil profile potential flow theory[J]. Journal of Wuhan University of Technology, 1980, (1): 42-56.

[67] Zhang S S. The method to solve plane potential flow by physical prolongation transformation relation[J]. Journal of Wuhan University of Technology, 1981, (3): 97-105.

[68] Zhang Z Y, Zhao P, Li Y F, et al. Wind Energy&Power Technology(Version 2)[M]. Beijing: Chemical Industry Press, 2010.

[69] Jiang H B. Lift performance of wind turbine with blade tip loss[C]. Proceedings of 2014 the 4th International Conference on Mechatronics and Intelligent Materials. Lijiang, 2014.

[70] Jin J T, Peng C Y, Pan L J, et al. Aerodynamic shape parameter calculation and 3D modeling approach for blade of large wind turbine[J]. Mechanical Design, 2010, 27(5): 11-13.

[71] Yang T, Li W, Zhang D D. Aerodynamic shape design and 3D solid model building study for blade of wind turbine[J]. Mechanical Design and Manufacture, 2010, (7): 190-191.

[72] Jiang H B, Cheng Z Q, Zhao Y P. Discussion on mathematical model of blade of wind turbine and blade function design method[J]. Mechanical Design, 2014, 31(5): 78-82.

[73] Xu X Y, Zhong T Y. Structure and matlab realization of cubic spline interpolation function[J]. Ordnance Industry Automation, 2006, 25(11): 76-78.

[74] Xu B Q, Tian D, Zhao D P, et al. Application of cubic spline interpolation in wind turbine blade design[J]. Journal of Inner Mongolia University of Technology(Natural science edition), 2010, 29(4): 179-183.

[75] Zhao W L. Aerodynamic Design and Flow Control Study for Large Wind Turbine[M]. Beijing: China Water & Power Press, 2013.

Appendices

Appendix Ⅰ Key terms interpretation

Flat airfoil: refers to the airfoil whose thickness and curvature are close to 0.

Function airfoil: refers to the airfoil generated by analytical function.

Design condition: refers to stable operation state or optimal operation state of wind turbine.

Optimal attack angle: attack angle with highest blade element efficiency (it has been proved to be attack angle for highest airfoil lift drag ratio).

Ideal twist: assume airfoil is unchanged along span, the difference between the blade inflow angle and optimal attack angle under design condition.

Ideal chord length: assume airfoil is unchanged along span and twist is ideal one, blade chord distribution along span obtained blade element-momentum theorem under design condition.

Relative radius: relative distance (to blade length) that a certain point on the blade to center of rotation.

Ideal blade: blade in airfoil structure that lift drag ratio is infinitely great under ideal twist, chord and fluid.

Ideal wind turbine: wind rotor composed by infinite number of ideal blades.

Practical wind turbine: wind rotor composed by limited number of practical blades.

Tip speed ratio associated extremity: when drag is 0 (or lift draft ratio is infinitely great), expression of power, torque, lift and thrust coefficient with highest performance (only function for tip speed ratio).

Function blade: refers to blade generated by analytical formula or parametric equation.

Smooth transition method: fairing method at the junction of different airfoils along span and junction between blade root cylinder and airfoil. It is also used for fairing connection among different parameters.

Sine curve fairing method: fairing method by about half cycle of sine curve for two unequal parameters, slope at the highest point or lowest point of sine curve for the junction should be equal to curve slope outside parameters connected.

Pressure distribution ring view: observe airfoil from airfoil circumcircle to the direction that is perpendicular to chord, and take anti-clockwise azimuth from trailing edge θ as independent variable, take $(C/2)\cos\theta$ (C refers to chord) x-coordinate of airfoil view point, and pressure as y-coordinate to obtain pressure distribution chart, which can describe pressure distribution of any point on airfoil, including pressure distribution nearby leading edge.

Appendix II Meanings of main symbols

R Blade length, i.e., the distance from blade tip to rotation axis center of wind turbine.

r The distance from blade span micro-segment dr to rotation axis center. Relative chord r/R is generally expressed as x.

C Cord length at r.

a Axial speed induction factor.

U Absolute wind velocity from infinite axial inflow of wind turbine.

u Wind velocity of wind turbine axis passes wind rotor.

b Tangential speed induced factor.

W Reverse relative wind speed caused by peripherad movement at blade r, which is equal to linear velocity at position r.

w Resultant velocity of reverse relative wind speed W caused by blade peripherad movement and tangent induced velocity bW.

v Synthetic air speed of u and w.

L Lift caused by wind velocity v, the direction is perpendicular to v, circumferential component is L_w, axial component is L_u.

C_L Lift coefficient.

D Drag caused by wind velocity v, the direction is the same as v, circumferential component is D_w, axial component is D_u.

C_D Drag coefficient.

C_p Press coefficient of airfoil.

C_P Power coefficient of blade or wind turbine.

C_M Torque coefficient of blade or wind turbine.

C_F Lift coefficient of blade or wind turbine.
C_T Twist coefficient of blade or wind turbine.
ζ Ratio between airfoil lift coefficient C_L and drag coefficient C_D, is known as lift-drag ratio.
φ Included angle for resultant wind velocity v rotation plane is called as incoming angle or inflow angle.
α Included angle of blade airfoil and resultant wind velocity v is actual attack angle of blade.
β Twist, refers to included angle of airfoil chord and rotation plane.
ω Blade rotation angular velocity.
λ Specific value of tangential linear velocity W at the position r and absolute wind velocity of infinite incoming flow U is called as linear speed velocity ratio.
λ_t Blade top linear speed ratio is abbreviated as tip speed ratio.
ρ Air density.
B Number of blades.
x Relative coordinate along span (take the direction from rotor to blade tip as positive, $x = r/R$).
y Relative coordinate of rotor (taking along-wind as positive).
z Relative coordinate of airfoil or blade surface is perpendicular to x and y direction.

Appendix III Common relation index

$$\bar{a} = \frac{1}{3} + \frac{1}{3}f - \frac{1}{3}\sqrt{1-f+f^2} \tag{9.2}$$

$$a_B = \bar{a}/f \tag{9.4}$$

$$\bar{b} = \frac{\bar{a}(1-\bar{a}/f)}{\lambda_t^2 x^2} \tag{9.3}$$

$$b_B = \bar{b}/f = \frac{\bar{a}(1-\bar{a}/f)}{\lambda_t^2 x^2 f} \tag{9.5}$$

$$\frac{C}{R} = \frac{16\pi}{9B} \frac{r}{R} \frac{1}{\left[\left(\lambda + \frac{2}{9\lambda}\right)C_L + \frac{2}{3}C_D\right]\sqrt{\left(\lambda + \frac{2}{9\lambda}\right)^2 + \left(\frac{2}{3}\right)^2}} \quad (3.17)$$

$$\frac{C}{R} = \frac{8\pi}{B} \frac{\bar{a}(1-\bar{a})x}{\left[\left(\lambda_t x + \frac{\bar{a}g}{\lambda_t xf}\right)C_L + gC_D\right]\sqrt{\left(\lambda_t x + \frac{\bar{a}g}{\lambda_t xf}\right)^2 + g^2}} \quad (9.13)$$

$$C_D = 2C_f + 2\sin^2\alpha = 2C_f + 2\left(\frac{C_L}{2\pi}\right)^2 \quad (6.27)$$

$$f = \frac{2}{\pi}\arccos\left\{\exp\left[-\frac{B(1-x)}{2x}\sqrt{1+\frac{\lambda_t^2 x^2}{(1-a)^2}}\right]\right\} \quad (9.1)$$

$$g = 1 - \bar{a}/f \quad (9.6)$$

$$\beta = \varphi - \alpha_b \quad (3.13)$$

$$\zeta = \frac{C_L(\alpha)}{C_D(\alpha)} \quad (3.9)$$